Procedures for Commercial Building Energy Audits

Second Edition

This publication was prepared under the auspices of ASHRAE Technical Committee 7.6, Building Energy Performance.

Principal Authors

Michael Deru,
National Renewable Energy Lab
TC 7.6 Chair

Jim Kelsey,
kW Engineering
Committee Co-Chair

Dick Pearson,
Pearson Engineering
Committee Co-Chair

Authors

Bruce Hunn,
Consultant

Kendra Tupper,
Rocky Mountain Institute

Dakers Gowans,
Left Fork Energy, Inc.

Megan Van Wieren,
Eaton Corporation

David Eldridge,
Grumman/Butkus Associates

Michael Levinson,
Enermodal

Keith I. Emerson,
Tri-State Generation and Transmission Association

Michele Friedrich,
PECI

Steve Carlson,
CDH Energy

Contributors

Adrienne Thomle,
Honeywell

Greg Towsley,
GRUNDFOS Pumps Corporation

Bernt Askildsen,
Lloyd's Systems

Mark Case,
ETC Group, LLC

Dave Moser,
PECI

Rob Risley,
FPL Energy Services

Eric Harrington,
Rocky Mountain Institute

Ron Nelson,
Institute for Market Transformation

Fredric Goldner,
Energy Management and Research Associates

Stanton Stafford,
Newcomb & Boyd

Updates and errata for this publication will be posted on the ASHRAE website at www.ashrae.org/publicationupdates.

Procedures for Commercial Building Energy Audits

Second Edition

ASHRAE

Peachtree Corners

ISBN 978-1-936504-09-1

180 Technology Parkway
Peachtree Corners, GA 30092
www.ashrae.org

Printed in the United States of America

Cover design by Tracy Becker, Graphics Specialist.

Library of Congress Cataloging-in-Publication Data

Procedures for commercial building energy audits. -- 2nd ed.
p. cm.
Prepared by Michael Deru, Jim Kelsey, and others.
"This publication was prepared under the auspices of ASHRAE Technical Committee 7.6, Building Energy Performance."
Includes bibliographical references.
Summary: "Provides guide for building owners, managers, and government entities on what to expect from an audit, building a team, levels of audit, writing audit report, analytical methods, approaches to site visits, on-site measurements, economic evaluation, best practices for auditors, analysis templates, and forms for field collection of data"-- Provided by publisher.
ISBN 978-1-936504-09-1 (softcover : alk. paper)
1. Commercial buildings--Energy conservation. 2. Commercial buildings--Energy consumption. 3. Commercial buildings--Cost of operation. I. Deru, Michael P. II. Kelsey, Jim, P.E. III. ASHRAE Technical Committee 7.6, Building Energy Performance.
TJ163.5.B84P75 2011
696--dc23

2011034908

ASHRAE Staff

Special Publications

Mark Owen
Editor/Group Manager of Handbook and Special Publications
Cindy Sheffield Michaels
Managing Editor
James Madison Walker
Associate Editor
Elisabeth Parrish
Assistant Editor
Meaghan O'Neil
Editorial Assistant
Michshell Phillips
Editorial Coordinator

Publishing Services

David Soltis
Group Manager of Publishing Services and Electronic Communications
Tracy Becker
Graphics Specialist
Jayne Jackson
Publication Traffic Administrator

Publisher

W. Stephen Comstock

Contents

Part 1—Defining the Levels of Effort of Commercial Building Energy Audits

Part 2—Best Practices for Conducting Energy Audits

Also included is online access to sample data collection forms and templates in Microsoft® Excel® spreadsheet format that may be downloaded and used as the bases for data collection and reporting results from energy audits and pdfs of an energy auditor checklist and a list of energy efficiency measures (EEMs) to consider when performing an energy audit.

List of Sidebars

Acknowledgments

This document is a much-expanded version of the first edition of *Procedures for Commercial Building Energy Audits* published in 2004 by ASHRAE. It represents the combined knowledge and expertise of many individuals. Through the leadership of the co-chairs of the committee for this book revision, Jim Kelsey and Dick Pearson, and the vision of the previous chair of ASHRAE Technical Committee 7.6, Adrienne Thomle, the working group has produced a foundational document for energy audits of commercial buildings.

The primary authors and reviewers of this document include Steve Carlson, David Eldridge, Keith Emerson, Michele Friedrich, Dakers Gowans, Bruce Hunn, Michael Levinson, Kendra Tupper, and Megan Van Wieren. Reviews and supporting content were provided by Adrienne Thomle, Bernt Askildsen, Dave Moser, Ron Nelson, Rob Risley, Mark Case, Eric Harrington, Fredric Goldner, Stanton Stafford, and Greg Towsley.

Significant material was contributed by Alexander Zhivov from *IEA ECBCS Annex 46: Energy & Process Assessment Protocol* (IEA 2009). Rocky Mountain Institute (RMI) did the preliminary development of the energy audit forms found in this publication.

Michael Deru
Chair, ASHRAE Technical Committee 7.6

March 2011

Preface—How to Use this Book

The goals of this edition of *Procedures for Commercial Building Energy Audits* are to

- define levels of effort for energy audits;
- provide a reference guide for building owners, managers, government entities, and other consumers illustrating best practices for conducting energy assessments and the associated deliverables; and
- serve as an introductory guide to best practices for energy auditors.

The intended scope of this volume is for existing commercial and institutional buildings. However, many of the procedures herein will be applicable to industrial or multifamily residential facilities as well.

Part 1 of this book discusses Level 1, Level 2, and Level 3 energy audits as defined by ASHRAE. The procedures outlined are the minimum requirements for the levels. These definitions have changed only slightly from the first edition of this book.

Part 2 outlines recommended audit procedures and best practices. This section is intended only as a reference and does not define required audit procedures. The intent of this section is to provide guidance to auditors and those who would like to hire professional energy assessors.

Part 3 includes additional useful resources, such as sample data collection forms and templates, unit conversions, a list of abbreviations and acronyms, and a references section.

The sample data collection forms and templates included in Part 3 may be downloaded and used as the bases for data collection and reporting results from energy audits from **www.ashrae.org/PCBEA**. The forms and templates are in Microsoft Excel® spreadsheet format. The Web site also includes pdfs of an energy auditor checklist reprinted with permission from Washington State University and a list of energy efficiency measures (EEMs) to consider when performing an energy audit.

Energy Audit Sample Forms and Templates

Gathering the right data with the appropriate amount of detail is a key component to achieving the maximum benefits of an energy audit. However, the process for audit data collection is not standardized and, often, crucial data may be overlooked.

To facilitate the data collection effort, sample energy audit forms and templates are provided in Part 3 and at www.ashrae.org/PCBEA. These include data collection forms for all components of a commercial building that contribute to heating and cooling loads or consume energy and are organized by building system (e.g., glazing, cooling towers, etc) and template procedures for completing the Preliminary Energy-Use Analysis (PEA), end-use breakdowns, and EEM summary tables. The purpose of these sample forms and templates is to assist building energy auditors in collecting the data required to complete comprehensive energy and financial analyses of proposed modifications to a building.

How to Use The Forms and Templates

The data collection forms, which are best suited to Level 2 and Level 3 audits, may be used to aid the auditor in gathering data at the site. They are very comprehensive and are intended to support the collection of data needed for a detailed audit of a facility. It may not be necessary to collect all the data covered by the forms for every facility; therefore, auditors will need to use their judgment as to what data is and what level of investigation are appropriate for each specific site. Auditors can use the following guidelines to determine which forms to use:

- Identify which information listed in the sample forms is needed for a specific project.
- Identify the anticipated source for each piece of data. Typical data sources include
  - facility managers,
  - as-built drawing sets,
  - utility bills,
  - building automation system (BAS) trend data,
  - on-site spot measurements,
  - temporary logger data, and
  - occupant surveys.
- Collect and organize as much data as possible prior to a site visit.
- Print the sample forms (sometimes multiples of each) and take them to the site (or take the electronic versions).

Some of the forms contain "guidance" sample forms and show examples and potential data sources of information that should be collected.

The sample template procedures for completing the PEA, end-use breakdowns, and EEM summary tables include basic unit conversion calculations and financial calculations that are consistent with the guidelines in this book. See the following sample spreadsheets:

- The file "ASHRAE PEA Template.xls" includes a sample format for the PEA methods in the Preliminary Energy-Use Analysis sections of Part 1 and 2.

- The file "ASHRAE Sample End-Use Breakdown.xls" includes a sample end-use breakdown consistent with the methods of the Level 2—Energy Survey and Engineering Analysis section of Part 1 and the Energy End-Use and Cost Allocation section of Part 2.
- The file "ASHRAE EEM Summary Table.xls" is a sample EEM summary table for reporting energy audit results with common financial performance metrics and economics consistent with the Level 2—Energy Survey and Engineering Analysis section of Part 1 and the Audit Reports section of Part 2.

Note that the forms and templates available on the Web site may be revised from time to time, so users should refer to the Web site for the most up-to-date version of any form or template they wish to use for their audit projects. When a file is revised, it will be posted to www.ashrae.org/PCBEA with a new date at the end of the file name; this revised file will replace any versions previously available on the Web site.

The Term *Energy Audit*

The term *energy audit* is an established term within the industry. It is widely used and understood by many. Ultimately we decided to retain the word *audit* for this publication because of its wide industry usage and to maintain a consistent title for the book, which is referenced in multiple publications and standards. However, we have chosen to use the terms *energy audit* and *energy assessment* interchangeably throughout. Also in popular use are the terms *energy analysis*, *energy survey*, *energy evaluation*, *energy investigation*, *energy assessment*, and others. We have used these terms within this publication as well. Many of these terms have achieved some common use in the industry and they may be more descriptive than the term *energy audit*.

For many reasons, *assessment* is preferable to *audit*—*assessment* carries none of the negative connotations of the word *audit*, which is commonly associated with an involuntary investigation of finances or taxes, where the perceived goal is to uncover mistakes and assess monetary penalty, and is often seen as a precursor to some kind of enforcement action or punishment. If we are to gain market acceptance for energy assessment, such negative connotations are best avoided. *Assessment* also conveys more accurately the activity at hand—that is, assessing the current state of a building and making recommendations to improve its performance. Our hope is to move toward a wider use of *assessment* yet maintain reference to the term *audit*.

Similarly, throughout this book we've used the terms *auditor*, *analyst*, and *assessor* interchangeably.

Part 1

Defining the Levels of Effort of Commercial Building Energy Audits

Levels of Effort

Overview

A commercial building energy analysis can generally be classified into the following three levels of effort:

- Level 1—Walk-Through Analysis
- Level 2—Energy Survey Analysis
- Level 3—Detailed Analysis of Capital Intensive Modifications

In addition, there is a Preliminary Energy-Use Analysis (PEA), which is a prerequisite for any audit, and there are targeted audits, which do not have a strictly defined level of effort bu t may be useful or necessary for some situations.

This section contains a brief overview of the PEA, each audit level, and targeted audits. More rigorous definitions for each level also follow in this section.

Preliminary Energy-Use Analysis (PEA)

The PEA precedes an audit of the building. During the PEA the analyst analyzes the historic utility use, peak demand, and cost; develops the Energy Cost Index (ECI) of the building (expressed in dollars per floor area per year); and develops the Energy Utilization Index (EUI) of the building (expressed in kBtu/ft^2 [MJ/m^2] per year). The analyst then compares the building EUI to similar buildings' EUIs to assess the potential for improved energy performance and to determine whether further engineering study and analyses are likely to produce significant energy savings. Monthly energy use and peak demand or, if available, interval billing data (such as 15 minute data), are reviewed to identify efficiency or behavioral modification opportunities.

Use the PEA Template found in Part 3 and at www.ashrae.org/PCBEA to assist with developing and comparing the ECI and EUI and analyzing monthly energy use.

A PEA provides necessary background data for Levels 1, 2, or 3 analyses.

Level 1—Walk-Through Survey

First, the building's energy cost and efficiency are assessed by analyzing energy bills, compiled in the PEA, and conducting a brief on-site survey of the building. A Level 1 energy survey will identify low-cost/no-cost measures for improving energy efficiency and provide a listing of potential capital improvements that merit further consideration. Because calculations at this level are minimal, savings and costs are approximate.

A Level 1 analysis is applicable when the desire is to establish the general energy savings potential of a building or to establish which buildings in a portfolio have the greatest potential savings. Level 1 results can be used to develop a priority list for conducting Level 2 and 3 audits.

Level 2—Energy Survey and Analysis

A Level 2 audit involves a more detailed building survey, including energy consumption and peak demand analysis. A breakdown of energy end uses within the building is developed.

A Level 2 energy analysis will identify and provide the savings and cost analyses of all practical energy efficiency measures (EEMs) that meet the owner's/operator's constraints and economic criteria, along with proposed changes to operation and maintenance (O&M) procedures. It may also provide a listing of potential capital-intensive improvements that require more thorough data collection and engineering analysis as well as an assessment of potential costs and savings. This level of analysis will provide adequate information for the owner/operator to act upon recommendations for most buildings and for most measures.

Level 3—Detailed Analysis of Capital-Intensive Modifications

The third level of engineering analysis focuses on potential capital-intensive projects identified during a Level 2 analysis. It requires more detailed field data gathering as well as more rigorous engineering and economic analyses, often including modeling (simulation) of the annual energy performance of the building and vendor pricing. It provides detailed project cost and savings calculations with a high level of confidence sufficient for major capital investment decisions. It often goes beyond the economic analysis of a Level 2 audit and uses a comprehensive life-cycle cost analysis (LCCA) as a decision-making tool.

Targeted Audits

A targeted audit is an investigation with a limited scope, typically a single energy-using system, central plant, or area of the building. Some examples include lighting-only audits, cooling tower replacements, boiler control assessments, and tenant improvement projects.

A targeted audit will identify and provide savings and cost analyses for retrofits and control strategy improvements for the systems of interest. The level of effort may be tailored to the needs of the facility. Because the audit is limited to a portion of the building, whole-building approaches such as end-use allocation and comparisons with historical utility bills cannot be used to provide a check on analytical methods. Similarly, whole-building simulation is typically not a cost-effective approach except for large systems. For this reason, targeted audits rely on measurements, data logging, and trend data to provide a check on the energy use estimates in the base case.

Discussion

There are not sharp boundaries between these levels. The levels are general categories for identifying the type of information that can be expected and an indication of the level of confidence in the results. While performing an energy analysis in a particular building, various measures may be subjected to different levels of analysis based on their savings potential or expected implementation costs.

The matrix presented in Table 1 is a general representation of the tasks required for each level of effort.

Table 1 — Energy Audit Required Tasks

Process	Level 1	Level 2	Level 3
Conduct PEA	•	•	•
Conduct walk-through survey	•	•	•
Identify low-cost/no-cost recommendations	•	•	•
Identify capital improvements	•	•	•
Review mechanical and electrical (M&E) design and condition and O&M practices		•	•
Measure key parameters		•	•
Analyze capital measures (savings and costs, including interactions)		•	•
Meet with owner/operators to review recommendations		•	•
Conduct additional testing/monitoring			•
Perform detailed system modeling			•
Provide schematic layouts for recommendations			•
Report	**Level 1**	**Level 2**	**Level 3**
Estimate savings from utility rate change	•	•	•
Compare EUI to EUIs of similar sites	•	•	•
Summarize utility data	•	•	•
Estimate savings if EUI were to meet target	•	•	•
Estimate low-cost/no-cost savings		•	•
Calculate detailed end-use breakdown		•	•
Estimate capital project costs and savings		•	•
Complete building description and equipment inventory		•	•
Document general description of considered measures		•	•
Recommend measurement and verification (M&V) method		•	•
Perform financial analysis of recommended EEMs		•	•
Write detailed description of recommended measures			•
Compile detailed EEM cost estimates			•

For ease of reading, an executive summary should be the first section of an energy analysis report. Some readers of such a report may be unable to comprehend the technical analyses involved, while others may demand a full presentation of the analyses for critique. Consequently, technical material should be presented in an appendix to the report, while the body of the report should guide the reader through the technical material and summarize the findings.

Information presented in this book outlines the engineering procedures that should be followed while performing an energy analysis. It is assumed that the analyst is a knowledgeable and competent individual. No attempt is made in this publication to prescribe specific methods of data gathering or data analysis, although recommendations for best practices are described in Part 2 of this book.

To assist with the organization of the data collected and the calculation procedures, this publication contains guideline forms that suggest the types of data to be gathered and their organization. It is recommended that the analyst develop and use appropriate data collection and organization forms specific to the size and type of the building(s) being analyzed.

The engineering services performed and the reports provided after performing Level 1, 2, and 3 analyses are progressively more specific and complex. The descriptions provided in Figure 1 are intended for the work and deliverables required in each level to build upon the previous level.

Preliminary Energy-Use Analysis (PEA)

Before any level of energy analysis has begun, it is valuable to perform a PEA to determine a building's current energy and cost efficiencies relative to other, similar buildings. This is normally done by calculating the energy use and cost per unit area per year, which can indicate the potential value of further levels of analysis. This preliminary analysis generally includes the following steps:

1. Determine the building's gross conditioned floor area and record this on the basic building characteristics form (use the PEA Template in Part 3 and at www.ashrae.org/PCBEA). Classify the primary use of the building. Ensure that the standard definition of *gross floor area* is used, which is:

    > the sum of the floor areas of all the spaces within the building with no deductions for floor penetrations other than atria. It is measured from the exterior faces of exterior walls or from the centerline of walls separating buildings, but it **excludes** covered walkways, open roofed-over areas, porches and similar spaces, pipe trenches, exterior terraces or steps, roof overhangs, parking garages, surface parking, and similar features.

2. Assemble copies of utility bills and summarize them for at least a one-year period, preferably for a two- or three-year period. Review the monthly bills for opportunities to lower costs by taking advantage of

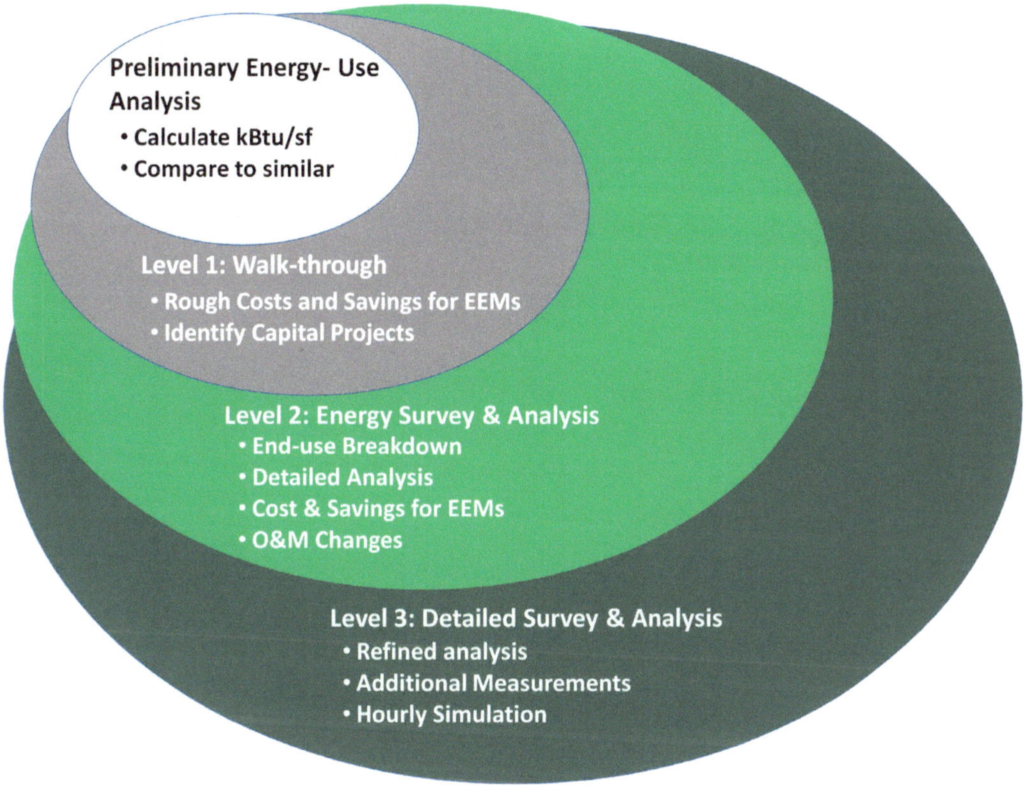

Figure 1 — Relationships of ASHRAE Energy Audit Levels 1, 2, and 3

different utility rate classes, taking into account peak electric demand patterns. Review the monthly patterns for irregularities. Note if a bill is missing or if it is estimated rather than an actual consumption value.

3. Complete the energy performance summary to develop the EUI and the ECI for each fuel or demand type and their combined total using methods outlined in ASHRAE Standard 105 (ASHRAE 2007a).
4. Compare the EUI and ECI with those of buildings having similar characteristics. A common benchmark comparison for peer buildings is the ENERGY STAR Portfolio Manager of the U.S. Environmental Protection Agency (EPA) (2011a). The owner/operator of the subject building may have similar buildings for this comparison. Comparison should also be made with publicly available energy indices of similar buildings. In all cases, care should be taken to ensure that comparison is made with current data, using consistent definitions of building usage and floor area.
5. Derive target energy, demand, and cost indices for a building with the same characteristics as the building being analyzed. A range of methods are available for this work, including
    - choosing from any database of similar buildings those buildings with the lowest energy index and

- choosing an index based on the knowledge of an energy analyst experienced with this type of building.

6. Compare the energy and cost savings for each fuel type if the building were to reach the target EUI. Using these values, determine whether further engineering analysis is recommended.

Use the PEA Template found in Part 3 and at www.ashrae.org/PCBEA to assist with developing and comparing the ECI and EUI and analyzing monthly energy use.

Level 1—Walk-Through Analysis

The Level 1 process includes all of the work performed for the PEA plus the following steps:

1. Perform a brief walk-through survey of the facility to become familiar with its construction, equipment, operation, and maintenance.
2. Meet with the owner/operator and occupants to learn of special problems or planned improvements (e.g., HVAC equipment replacements, aesthetic upgrades, etc.) of the facility and any operation or maintenance issues. Determine whether any maintenance problems and/or practices affect efficiency.
3. Perform a space function analysis, guided by sheet 1.12 in the PCBEA Sample Forms in Part 3 and at www.ashrae.org/PCBEA. Determine whether efficiency may be affected by functions that differ from the original functional intent of the building.
4. Identify low-cost/no-cost changes to the facility or to O&M procedures and estimate the approximate savings that will result from these changes.
5. Identify potential capital improvements for further study and provide an initial rough estimate of potential costs and savings.

The report for a Level 1 analysis should contain the building characteristics and energy use summary as well as the following items:

- Quantification of any savings potential from changing to a different utility rate structure.
- Discussion of irregularities found in the monthly energy use patterns, with suggestions about their possible causes.
- The EUIs of similar buildings. Report the source, size, and date of the sample used in this comparison. The names of comparable buildings should be given if known.
- The target EUI and the method used to develop the target index. Where comparison is made to other buildings, state their names or the source of the database. Where the experience of someone other than the report author is used to develop the target, provide the source. Where the target is developed by calculation, show the calculation or

quote the name and version of the software used and include both input and output data.

- Total energy and demand costs by fuel type for the latest year and preceding two years, if available. Show potential savings in dollars using the energy index format of ASHRAE Standard 105 (ASHRAE 2007a).
- The fraction of current costs that would be saved if the energy index were brought to the target level.
- A summary of any special problems or needs identified during the walk-through survey, including possible revisions to O&M procedures.
- A listing of low-cost/no-cost changes with estimated savings for these improvements.
- The potential capital improvements, with an initial rough estimate of potential costs and savings

Level 2—Energy Survey and Engineering Analysis

The Level 2 analytical procedure is guided by Level 1 analysis and includes the following additional steps:

1. Review mechanical and electrical system design, installed condition, maintenance practices, and operating methods.
2. Describe and analyze the energy-using systems of the building, resulting from on-site observation, measurement, and engineering calculations. Use the PCBEA Sample Forms found in Part 3 and at www.ashrae.org/PCBEA to document the building systems, including the following systems:
    - Envelope
    - Lighting
    - Plug loads
    - HVAC
    - Domestic hot water
    - Laundry
    - Food preparation
    - Refrigeration
    - Conveying
    - Pools/saunas/spas
    - Process loads
    - Others
3. Review existing O&M problems and logs. Review planned building changes or improvements and estimate their costs.
4. Measure key operating parameters and compare them to design levels, for example, operating schedules, heating/cooling water temperatures, the supply air temperature, the space temperature and humidity, ventilation quantities, and task lighting levels. Such measurements may be taken on a spot basis or logged manually or electronically.

5. Prepare a breakdown of the total annual energy use into end-use components, as illustrated in Part 2 of this book in the section Energy End-Use and Cost Allocation. A number of calculation methods are available, ranging from simplified manual calculations to fully detailed computer simulation of hour-by-hour building operations for a full year.
6. List all possible modifications to equipment and operations that will save energy. Select those that might be considered practical by the owner/operator. Perform preliminary cost and savings estimates.
7. Review the list of practical modifications with the owner/operator and select those that will be analyzed further. Prioritize the modifications in the anticipated order of implementation.
8. Create integrated bundles of measures where successive efficiency measures have significant interactive effects. Measures should be packaged together to achieve different levels of whole-building energy use reduction and financial performance. Package measures together to reduce loads to lower equipment capital costs and increase interactive energy savings. Capture interactions in at least one package that meets the project's financial criteria or savings goals.
9. For each practical measure, estimate the potential savings in energy costs and the building EUI. To account for interaction between modifications, assume that modifications with the highest operational priority and/or best return on investment will be implemented first. A number of calculation methods are available, ranging from simplified manual calculations to iterative computer simulations.
10. Estimate the implementation cost of each practical measure.
11. Estimate the impact of each practical measure on building operations, maintenance costs, and nonenergy operating costs.
12. Estimate the combined energy savings from implementing recommended bundles of measures and compare the estimated savings to the potential savings derived in the Level 1 analysis. It should be clearly stated that savings from each modification are based on the assumption that all previous modifications have already been implemented and that the total savings account for all of the interactions between the modifications.
13. Prepare a financial evaluation of the estimated total potential investment using the owner's/operator's chosen techniques and criteria. These evaluations may be performed for each practical measure or for combinations of practical measures.
14. During the development of the report of the Level 2 analysis, meet with the owner/operator to discuss priorities and to help select measures for implementation or further analysis.

The report for a Level 2 analysis should contain at least the following items:

- A summary of the current energy use and cost associated with each end use. Show calculations performed or quote the name and version of software used and include both input and output data. For ease of reading, present end-use data in pie charts or other graphic formats. Further details and examples are presented in the Billing Data Review section of Part 2 of this book. Provide interpretation of differences between actual total energy use and calculated or simulated end-use totals.
- A description of the building, including typical floor plans and inventories of major energy-using equipment. (This information may be included as an appendix.)
- For each practical measure, provide
    - a discussion of the existing situation and how excess energy is being consumed;
    - a description of the measure, including its impact on occupant health, comfort, and safety;
    - a description of any repairs that are required for a measure to be effective;
    - the impact of the measure on occupant service capabilities, such as ventilation for late occupancy or year-round cooling;
    - an outline of the impact of the measure on operating procedures, maintenance procedures, and costs;
    - the expected life of new equipment and the impact on the life of existing equipment;
    - an outline of any new skills required of operating staff and training or hiring recommendations;
    - calculations performed, or the name and version of the software used (including both input and output data); and
    - nonenergy benefits, especially improvements to health, safety, and environment, and decreases in equipment runtime and labor hours.
- A table listing the estimated costs for all practical measures and recommended bundles of measures, the savings, and a financial performance indicator (for example, simple payback period). For the cost of each measure, show the estimated accuracy of the value quoted. This table should spell out the assumed sequence of implementation and state that savings may be quite different if a different implementation sequence is followed.
- A list of measures considered but deemed to be impractical, with brief reasons for rejecting each.
- A discussion of any differences between the savings projected in this analysis versus any savings projected in a prior Level 1 analysis.
- Overall project economic evaluation.

- Recommended measurement and verification methods that will be required to determine the actual effectiveness of the recommended measures.
- Discussion of feasible capital-intensive measures that may require a Level 3 analysis.

Level 3—Detailed Analysis of Capital-Intensive Modifications

The Level 3 analytical procedure is guided by Level 1 and 2 analyses and the owner's/operator's selection of measures for greater definition. It follows Level 1 and 2 work and also includes the following additional steps:

1. Expand the definitions of all modifications requiring further analysis. Consider system interactions to create integrated packages of recommendations.
2. Review measurement methods and perform additional testing and monitoring as required to allow determination of feasibility.
3. Perform accurate modeling of proposed modifications. Ensure that modeling includes system interaction.
4. Prepare a schematic layout of each of the modifications.
5. Estimate the cost and savings of each modification and each integrated bundle of modifications. Perform a LCCA to inform decision making.
6. Meet with the owner/operator to discuss/develop recommendations.

The report for a Level 3 analysis should include the following items, as a minimum:

- The text, schematics, equipment lists, and manufacturer's cut sheets necessary to completely describe all proposed changes to physical equipment. Matters of a final design nature may be left to subsequent engineering as long as the cost of such engineering is included in the budget. Firm price contractor quotations for key parts of any measure may be included. Cost estimates should show contingencies separately and report the expected accuracy of the budget.
- A description of system interactions and explanations of why certain efficiency measures should be bundled together.
- A financial evaluation of the estimated capital investment and projected savings. Use LCCA along with the owner's/operator's chosen techniques and criteria.

Targeted Audits

A targeted audit is an energy assessment of a specific system or end use at a facility. Whereas the Level 1, 2, and 3 audits provide a comprehensive approach to energy savings, a targeted audit focuses more narrowly on an area of particular interest to a customer.

For example, in a building with a chiller at the end of its useful life, an owner may want to take advantage of the need for replacement to investigate efficiency options, review sizing specifications, and examine alternative operating strategies. Given the owner's system-specific focus, a full audit would be excessive and costly, so an energy assessment provider could recommend a targeted audit to study only the chiller replacement and control options.

Other common examples of systems where a targeted approach may be used include the following:

- Packaged equipment replacements
- Compressed air systems
- Building additions
- Lighting system upgrades

A targeted energy audit has the advantage of allowing a more in-depth approach to a specific area or system. Because the scope is limited, the auditor can focus on the details of the system under study. With a reduced scope, a more detailed study of the equipment or system can be achieved at lower cost compared to comprehensive energy audits.

The targeted audit approach provides unique challenges for the energy auditor. Because only a specific area or end use is under study, an end-use breakdown or energy balance of the facility is not an option, so it may be more difficult to attribute the correct portion of the utility bill to the system under study. This makes additional measurement and monitoring of the existing system particularly important to ensure an accurate picture of the base-case conditions, loads on the system, and existing control methods. Targeted approaches also typically cannot rely on whole-building hourly simulation tools (e.g., DOE-2 [JJH 2009a] and Energy Plus [EERE 2011a]) because the information needed to create these models involves the entire building and its systems and is therefore outside the scope of the investigation. Targeted audits commonly rely on engineering analysis performed via spreadsheets to estimate energy and demand requirements for different equipment and operating scenarios. For some technologies, specialized software may be used to supplement or replace simple engineering analysis, for example, the use of AirMaster+ (EERE 2011b) for air compressors.

Targeted Audit Methods

The following approaches will help ensure that energy and cost savings estimates for targeted audits are comparable to more comprehensive strategies:

- Clearly define the scope of the audit and the systems to be studied.
- Interview building O&M personnel to determine their opinions of how equipment is operated and maintained and whether there are other factors to be considered in any decision, such as special needs or maintenance issues.
- Reduce uncertainty in assumptions by using measurements and data logging.

- Use spot measurements to verify key inputs such as full-load power requirements and equipment efficiencies.
- Monitor performance over time to observe control sequences, system performance, and occupant behavior.
- Calibrate base-case energy use estimates to measured data when possible. This may require extrapolation of short-term data to a full year with proper accounting for weather and occupancy over time.
- Perform "reality checks" on end-use energy estimates. For instance, in a lighting-only audit, check whether the calculated energy use of the lighting systems as a percentage of billed energy use aligns with industry norms. Use end-use-specific averages for your building type when possible.

Part 2

Best Practices for Conducting Energy Audits

The Energy Audit Process

This section of the book provides guidance on best practices for conducting commercial building energy assessments. This guidance is informative only and not intended to alter the level-of-effort definitions of Part 1 of this book.

ASHRAE and other entities have defined *energy audits* as processes in which engineering service providers identify and recommend efficiency opportunities to clients. The objectives of an energy audit are to identify and develop measures that will reduce the energy use and/or cost of operating a building and/or will improve the indoor environmental quality experience by the occupants. The results should be presented in a format that provides complete information needed by an owner/operator to decide whether or not to implement the recommended measures. As a general overview, an energy audit includes the following steps:

1. Collect and analyze historical energy use.
2. Study the building and gather data on its operational characteristics and indoor environmental quality (use the PCBEA Sample Forms found in Part 3 and at www.ashrae.org/PCBEA).
3. Identify potential measures that will reduce the energy use and/or cost and/or will improve the occupants' indoor environmental quality.
4. Perform an engineering and economic analysis of potential modifications.
5. Prepare a list of appropriate measures and group measures into bundles that will capitalize on potential synergies and cost reductions.
6. Prepare a report to document the analysis process and results.

The energy audit may also include additional steps toward implementation of the recommended measures. These steps may include additional engineering and design, construction administration and management, commissioning (Cx), or measurement and verification (M&V) of energy savings.

Overview

A team approach helps ensure that an energy audit results in recommendations that can be implemented at the site. By involving multiple parties, the audit takes advantage of each team member's strengths. By actively involving building staff, the process can ensure that their concerns are heard, that their observations about the building add value, and that those who will be left to maintain the building are familiar with recommended equipment and controls. This section describes who should be involved in a building audit, their roles, and the requirements.

Selecting team participants depends on the type of project and building under consideration. Regardless of the project delivery method, the audit process will require experienced, committed, eager, and innovative minds carefully selected by the building owner. When selecting team members, it is important for the owner to define shared goals. While definitive quantitative or contractual goals should be explored once the entire team is selected, the building owner should develop high-level goals and communicate them to the project team as members are interviewed and selected.

An energy audit of a facility should provide sufficient information for the owner/operator and/or manager of a facility to understand the energy use characteristics of the building. When feasible, this analysis breaks down the total energy use and cost for the facility into various end uses, such as heating, air conditioning, lighting, etc., and shows the potential for energy and cost savings for each end use.

The engineering analysis should also provide the owner/operator with all the information needed to allocate necessary resources to reduce the building's energy use and/or cost. This includes outlining any changes in the facility's operation and maintenance (O&M), including different personnel requirements, as well as presenting an economic analysis of any capital improvement projects.

The engineering analyst is encouraged to follow a systematic approach in identifying, selecting, ranking, and grouping recommended measures. While it is useful to have a procedure to follow, consider factors beyond technical issues, such as the regulatory environment, financing options, and occupant requirements. An efficiency measure that is highly effective under some conditions may have little or no effect under others.

The key elements of a commercial building energy audit/analysis are described in Figure 2.

Different levels of energy audit can be performed on any given building, or group of buildings, providing information that may be used for widely varying purposes. A building owner contemplating major energy-saving capital improvements will need a significantly higher level of confidence in the analysis than an owner who simply wishes to compare the level of efficiency of the building to that of other, similar buildings.

As a result, the levels of effort have been organized into the following categories:

- Preliminary Energy-Use Analysis (PEA)
- Level 1—Walk-Through Analysis
- Level 2—Energy Survey and Analysis
- Level 3—Detailed Analysis of Capital-Intensive Modifications
- Targeted Audits

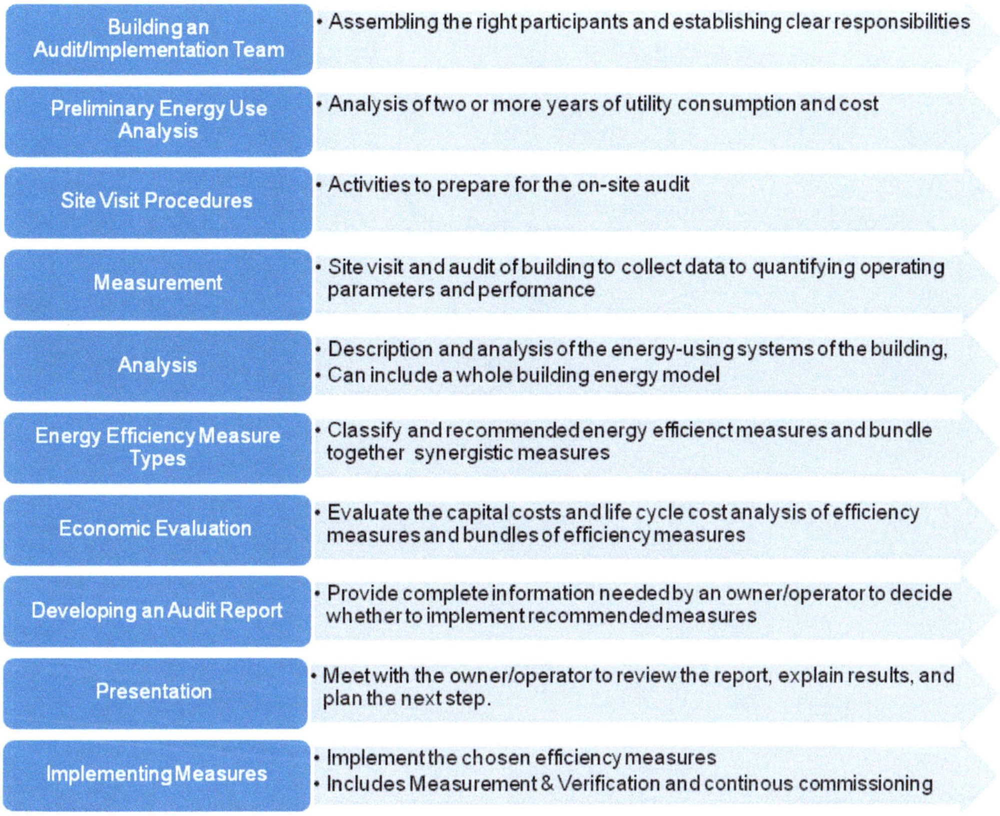

Figure 2 — Key Elements of the Audit Process

The different levels are defined in Part 1 of the book, along with the typical process of analysis and report contents for each level. Each succeeding level of analysis builds upon the previous level. A joint decision should be made by the building owner and energy analyst as to the level that is appropriate for the owner's needs.

Regardless of the level of audit being pursued, the key elements in the systematic approach remain the same. The following sections explore each key element in greater detail.

Building an Audit/Implementation Team

An energy audit is only as good as the information on which is it based. Needed data and expertise are often dispersed among many entities. Assembling the right participants, establishing clear responsibilities, and setting goals are key elements of an effective energy assessment. Building an audit team should focus on setting the project up for success.

Participants

A number of internal and external stakeholders and service providers can participate in an energy assessment. Depending on the facility type, size, and management structure, the composition of the team will vary from assessment

Build a Balanced Team

To make sure your energy-saving project will be a success, you'll need the following:

- Management with commitment to the project
- Financial staff who understand risks and rewards
- Trained building engineers who understand and can maintain the project
- Trusted contractors and vendors
- Utility account representatives to coordinate technical assistance and offer incentives
- Engaged and informed building occupants
- A trained and experienced energy assessment provider (auditor) who can identify, analyze, and present opportunities for energy savings

to assessment. However, there are certain entities that are common to many audits. These groups each have their own interests connected to the energy assessment project:

- Building Owner/Manager/Assessment Administrator
  - Primary interest is improving building energy efficiency at least cost.
  - Are concerned about the cost and effectiveness of the assessment work.
- O&M Staff
  - Primary interest is operating the site cost-effectively.
  - May also be concerned about improving the functionality and comfort of the building.
- Contractors
  - Primary interest is fulfilling their tasks cost-effectively.
  - May be concerned about maintaining long-term contracts with the building owner.
- Energy Assessment Provider
  - Primary interest is identifying energy efficiency measures (EEMs).
  - May also be concerned about accomplishing the assessment on time and on budget.
- Energy Service Companies (ESCOs)
  - Primary interest is identifying energy investment projects.
- Building Occupants/Tenants
  - Primary interest is having a quality built environment in which to perform their work.
  - May be concerned about ongoing energy costs.

- Utilities
  - Primary interest is providing energy according to their region's regulatory frameworks.
  - May have energy-efficiency goals and incentive programs for their service areas.

Each of these parties has specific skill sets and contributions to make regarding the energy assessment.

Building Owner

The building owner/manager may be an individual, a corporate entity, a government agency, a board of directors, or any number of other ownership/management structures. For the purposes of an energy assessment, ownership may be represented by a third party. The building owner is a necessary participant in any energy assessment (as the client) and is the ultimate decision maker regarding how the assessment's findings will be implemented. As such, the owner or manager has information critical to the energy assessment team, such as the following.

- The programmatic, conditioning, occupancy, scheduling, and other performance requirements of the building, as well as resource efficiency goals. These may be summarized in an Owner's Project Requirements document.
- The economic constraints surrounding the implementation of potential building improvements. The owner may have payback, return on investment, net present value, or other criteria that guide what will and won't be considered for implementation.

This information should be understood by the energy assessment provider before the audit begins.

The building owner also needs to direct site staff, contractors, and building occupants to cooperate with the energy assessment team during the audit process. This may include authorizing the use of various resources to support the audit, such as staff and contractor time, building information, and access.

Operations and Maintenance (O&M) Staff

The staff on site (energy managers, O&M personnel, etc.) may be employed directly by the building owner/manager or may be retained through a maintenance contract. In all cases, they should know what is expected of them before the assessment commences. They should also be informed of the assessment and report schedule for budgeting and long-term planning purposes.

Commitment from the site staff is needed well before the audit begins, preferably when the building is selected for assessment. The site staff should be made aware of resources, access, and information that will be needed to assist the assessment team before any site visits. Providing this information during the initial kick-off meeting is not recommended.

During the site visit, the assessment team needs assistance from people who know the building and the energy-using systems and can answer technical questions. Having someone with keys to the mechanical rooms is not enough. It is important to have a competent guide for each area to provide in-depth information. Critical information possessed by the O&M staff includes the following:

- Building information (utility billing data, plans, as-built drawings and specifications, O&M manuals and programs, an equipment database, building automation/control system specifics, etc.).
- Knowledge about the history and past issues associated with the performance of energy-consuming building systems, building comfort, and energy use.
- Recommendations for potential solutions to building problems and ideas to reduce energy and water use.
- Cost data for current O&M strategies.
- Site-specific cost issues, including process capacity utilization, material utilization, labor, waste, and equipment. Often, these costs greatly exceed the energy costs. Energy assessment teams need to understand how energy-efficiency recommendations might also reduce or inflate other building costs.

As the assessment proceeds, the O&M staff should be kept informed of the EEMs under consideration, especially those involving retrofits that might require different O&M skills, staffing, and/or training. Planning for such retrofits must include appropriate budgeting for staffing and training. This is especially important to ensure the persistence of energy savings. If O&M staff don't have the training or time to monitor advanced control strategies or equipment on an ongoing basis, then EEMs can prove ineffective. In some cases, the costs to maintain new systems can equal or exceed the resultant energy savings. Securing realistic O&M cost and capacity projections is essential.

Significant involvement from O&M staff leads to an improved energy assessment. Additionally, the O&M personnel's sense of ownership of the process can develop into a commitment to implementation.

Contractors

Contractors can play a number of roles in the operation and retrofit of a building (Figure 3). As mentioned previously, a contractor may be responsible for building system maintenance. Many facilities have a controls contractor who manages their building automation system (BAS). General contractors can be used to manage construction and retrofit projects, with mechanical, electrical, and other subcontractors hired to implement EEMs (see Figure 3). The relevant information possessed by contractors includes

- specific information about building and control system operation;
- cut sheets, specs, and other equipment-specific data; and
- estimates of potential equipment, installation, and maintenance costs.

Figure 3 — A Contractor Installing Task Lighting

It is important to manage contractor relationships in a manner that is transparent and free of any conflicts of interest. The building owner/manager should provide guidelines for contractor engagement.

Energy Assessment Provider

The energy assessment provider may be one person or a team with several members. The required expertise of the member professionals will vary according to the complexity and programmatic use of the site. Simple buildings may be audited by one expert examining all the energy-using systems as well as the building envelope, while larger and more complex buildings require collaboration among a variety of disciplines. The team usually includes the following:

- A building envelope expert or construction engineer
- A mechanical engineer (HVAC, hot water, etc.)
- An electrical systems expert (lighting, plug loads, etc.)
- A control systems expert
- An energy modeler experienced in estimating savings from both individual EEMs and bundles of EEMs with savings interactions; the modeler may use a spreadsheet-based approach or modeling software such as eQUEST (JJH 2009b)

Other specialists may also be needed in the areas of compressed air, refrigeration, food service, pools, laboratories, healthcare facilities, or production processes. The expertise of energy auditing does not depend on a strictly defined field of knowledge but a combination of skills and procedures including multiple disciplines. Assessment teams need to be able to understand the overall energy performance of a building, and how energy is consumed by specific end uses, with the ability to determine how efficiency performance can be improved. The more complex a building is and the more specialized its energy-using systems are, the more important it is to have experienced energy auditors.

It is also beneficial to include professionals with Cx and M&V expertise. This gives the team insight into developing specifications for EEMs that avoid installation problems. These professionals can also provide recommendations on how to best ensure the persistence of energy savings by providing the O&M staff with ongoing diagnostic energy use information.

Many auditing professionals come from engineering, architecture, science, and construction backgrounds. While there are many emerging certification programs aimed at validating energy auditing expertise, a common industry standard has yet to be established in the United States. Emerging certifications include:

- Building Energy Assessment Professional (BEAP) (offered by ASHRAE)
- Certified Energy Auditor (offered by Association of Energy Engineers [AEE])
- Certified Energy Manager (offered by AEE)
- High-Performance Building Design Professional (HBDP) (offered by ASHRAE)
- Certified Measurement and Verification Professional (CMVP) (offered by AEE and Efficiency Valuation Organization)

As with any service provider, past project experience, training, professional affiliations, and references should be examined before retaining services.

How to Hire an Energy Auditor

1. Determine what type of audit you are likely to need: Level 1, 2, or 3.
2. Determine if the energy auditor will play a role in the other phases of the project, e.g., contractor procurement, developing performance specifications, etc.
3. Use either sole-source contracting or competitive bidding.
   - Using sole-source contracting can result in a more quickly executed contract, but getting competitive bids may result in a more competitive price.
4. Look for the following in a good auditing firm:
   - That proposals include references, work samples, and resumes of key staff members.
   - That auditor staff have professional certifications (see the list of available certifications in the section Energy Assessment Provider)—consider whether you need general or specialized auditing experience, and consider firms with multidisciplinary competence for comprehensive audits or specialized expertise for targeted audits that involve specialized systems.
   - That the auditor is vendor-neutral to avoid bias and/or conflicts of interest.

Developing a detailed facility-specific scope of services will help ensure that a provider or providers is/are well suited to a particular audit.

Building Occupants/Tenants

Building occupants may be employees, commercial tenants, students, or any other population that uses the building on a regular basis. Usually this party plays a minor role in energy assessments. However, this group can offer useful information about building comfort and system performance requirements.

Building tenants play a larger role in the energy assessment when lease arrangements separate building system ownership and utility payments. In these cases, the legal nuances of the various parties' economic incentives to reduce energy use, such as lease terms and utility payments, need to be understood. Tenants may become partners with building owners regarding the assessment and subsequent implementation of EEMs and should not be overlooked.

Utilities

Energy is usually either purchased directly from a utility or through a commodity broker. When it is purchased from a utility, there may be incentive programs that can aid in the energy assessment and subsequent implementation of EEMs. For larger facilities, utilities will often assign a representative to help building owners utilize these programs. The utility should be contacted early in the assessment process to discover what resources are available.

Utilities and commodity brokers also dictate energy price and rate structures. This utility-specific information is essential to the accurate prediction of energy savings.

Project Management

There are many project management models that can be used to guide an energy assessment. In most cases, the prime responsibility for project management falls on the audit provider. This manager, in conjunction with the building owner/manager/energy assessment administrator, needs to clearly define the goals and general guidelines for the assessment activity and coordinate the budgeting, scheduling, and monitoring of the audit.

The project manager also needs to provide logistical coordination and ensure information (both incoming and outgoing) reaches the relevant parties. He or she is the main coordinator for site visits and reporting. The project manager is also responsible for quality control throughout the assessment. If there are several members on the assessment team, it is essential that the project manager make sure every member is aware of significant deadlines and associated deliverables.

Data management is critical for a successful energy assessment. The project manager must make sure that all preliminary data are available before the site visit, that all information needs are identified during the assessment, and that all information from the assessment team is included in the report. The project manager may have assistants for data collection or report compilation but must personally take all responsibility for incoming and outgoing information.

All team members should have a clear picture, before the assessment begins, of what is expected of them, both during site visits and in the reporting phase. The team and the project manager should have a clear understanding of the communication protocol surrounding contact with other parties.

Preliminary Energy-Use Analysis (Billing Data and Energy End-Use Analysis)

Analyzing historical utility data and preliminary building information can provide important insight into current building operation, energy end uses, and performance. This may assist in determining if an engineering study and analysis are likely to produce significant energy savings. Completing the PEA prior to conducting the site assessment will provide a basis for understanding where the potential energy and cost savings are within the building. It will also identify any anomalies in the utility bills, which may prompt the auditor to further investigate inconsistencies.

Use the PEA Template in Part 3 and at www.ashrae.org/PCBEA to assist with analyzing historical data and comparing key metrics, such as EUI, to various baselines and benchmarks. Having gained an understanding of baseline energy use and comparisons of energy use in similar buildings, the auditor can then establish target energy use and cost for the building being analyzed.

The PEA should determine any unusual monthly or annual energy use patterns, highlight which equipment uses the most energy, and identify efficiency or behavioral modification opportunities. The following sections provide a methodology for analyzing utility data to achieve an understanding of commercial building operation and performance.

Billing Data Review

Historical building utility data can usually be obtained from the owner or operator of the building before conducting the site assessment. Monthly energy use, power demand, and cost for at least the last 12 months should be collected. At least three years of data is necessary to sufficiently understand historical energy use profiles, identify anomalous months, and understand the effects from past changes to the building or its operation. Power demand data may be available for some buildings in 15-minute increments. To understand the costs that are associated with building energy use, it is important to obtain each rate structure from the utility provider(s).

Monthly and annual energy use from collected utility bills is most easily analyzed using tables or graphs. Refer to the PEA Template in Part 3 and at www.ashrae.org/PCBEA for templates of what information to collect and analyze. Bills for fuel consumption (i.e., natural gas, oil, electricity, etc.) should be compiled and examined to determine both the amount of energy used and the cost. Organizing this information in a central location, usually in tabular form, should be done first. Often, the data need to be time shifted to standard monthly values from the meter read dates or prorated to daily usage. This can be accomplished by multiplying weighted daily average energy use by the number of days per month. This data should then be put in to graphic form to examine the patterns and identify any anomalies. Graphs or charts of energy usage by fuel type by month or by average outdoor air temperature (normalized to energy use/day) provide valuable information. An example of compiled utility data is shown in Figure 4.

Bill #	Month	Year	Month/Yr	Days	Electric Demand Peak kW	Electric Use kWh	Electric Use kWh/day	Electric Cost $	Load Factor	kBtu
8	Jan	2006	Jan-06	34	163	87,680	2,579	$ 5,923	66%	299,252
9	Feb	2006	Feb-06	29	165	80,480	2,775	$ 5,483	70%	274,678
10	Mar	2006	Mar-06	29	165	79,360	2,737	$ 5,462	69%	270,856
11	Apr	2006	Apr-06	28	163	74,720	2,669	$ 5,249	68%	255,019
12	May	2006	May-06	29	165	79,520	2,742	$ 4,986	69%	271,402
13	Jun	2006	Jun-06	32	168	94,240	2,945	$ 5,368	73%	321,641
14	Jul	2006	Jul-06	31	173	91,840	2,963	$ 5,434	71%	313,450
15	Aug	2006	Aug-06	28	165	82,080	2,931	$ 5,013	74%	280,139
16	Sep	2006	Sep-06	32	163	88,000	2,750	$ 5,019	70%	300,344
17	Oct	2006	Oct-06	29	179	80,320	2,770	$ 4,880	64%	274,132
18	Nov	2006	Nov-06	29	173	76,800	2,648	$ 4,617	64%	262,118
19	Dec	2006	Dec-06	33	184	91,360	2,768	$ 5,206	63%	311,812
20	Jan	2007	Jan-07	36	179	95,680	2,658	$ 5,652	62%	326,556
21	Feb	2007	Feb-07	29	165	79,680	2,748	$ 5,096	69%	271,948
22	Mar	2007	Mar-07	29	173	77,440	2,670	$ 5,132	64%	264,303
23	Apr	2007	Apr-07	31	173	80,640	2,601	$ 5,107	63%	275,224
24	May	2007	May-07	30	160	79,680	2,656	$ 4,845	69%	271,948
25	Jun	2007	Jun-07	30	173	79,680	2,656	$ 5,199	64%	271,948
26	Jul	2007	Jul-07	31	166	86,880	2,803	$ 4,817	70%	296,521
27	Aug	2007	Aug-07	29	166	81,600	2,814	$ 4,502	71%	278,501
28	Sep	2007	Sep-07	31	160	83,840	2,705	$ 4,462	70%	286,146
29	Oct	2007	Oct-07	28	168	70,720	2,526	$ 4,575	63%	241,367
30	Nov	2007	Nov-07	31	155	73,920	2,385	$ 4,632	64%	252,289
31	Dec	2007	Dec-07	31	152	79,680	2,570	$ 4,776	70%	271,948
32	Jan	2008	Jan-08	33	163	85,600	2,594	$ 5,004	66%	292,153
33	Feb	2008	Feb-08	30	170	79,520	2,651	$ 4,867	65%	271,402
34	Mar	2008	Mar-08	28	160	73,600	2,629	$ 4,575	68%	251,197
35	Apr	2008	Apr-08	29	168	74,560	2,571	$ 5,363	64%	254,473
36	May	2008	May-08	34	162	83,360	2,452	$ 6,042	63%	284,508
37	Jun	2008	Jun-08	28	162	67,200	2,400	$ 5,488	62%	229,354
38	Jul	2008	Jul-08	33	165	3,520	107	$ 2,861	3%	12,014
39	Aug	2008	Aug-08	28	168	2,880	103	$ 2,884	3%	9,829
40	Sep	2008	Sep-08	31	150	37,280	1,203	$ 6,458	33%	127,237
41	Oct	2008	Oct-08	30	154	37,280	1,243	$ 3,749	34%	127,237
42	Nov	2008	Nov-08	30	154	69,920	2,331	$ 3,749	63%	238,637
43	Dec	2008	Dec-08	31	170	77,280	2,493	$ 5,425	61%	263,757

Figure 4 — Example of Compiled Utility Data

Monthly/Annual Energy Use

Once the utility data have been compiled, they can be modified into numerous graphs and charts for easy analysis. Understanding how the annual energy consumption is distributed between utility types is a good starting point, as this provides important information about the facility and what equipment should be targeted for energy use reductions. For example, Figure 5 indicates that electric energy accounts for more than half of the total building energy use. This energy use breakdown allows the auditor to focus the site assessment on equipment using electric energy and to identify potential efficiency opportunities.

Next, annual utility profiles should be graphed to provide an overview of the building operation. These graphs show how consumption varies from month to month and in comparison to previous years. Any unexpected pattern in the energy use raises the possibility for significant energy or cost savings by identifying and correcting that operation. Sometimes an anomaly on the graph reflects an error in billing, but generally the deviation shows that some equipment operation is not optimized and has gone unnoticed or is not understood by the owner or building operator.

Figure 6 provides an example of an expected annual natural gas utility profile. From this graph we can see that the natural gas use is lowest during hot summer months and peaks during periods of cold outdoor air temperatures.

Daily Energy Use

Daily power demand profiles provide information about building operation beyond the annual and monthly utility analysis discussed in the previous section. Graphs of 15-minute power demand indicate when peak building energy use occurs. Comparing this data to building equipment operation schedules, occupancy schedules, and weather data can be used to determine if and during what periods the potential for energy-efficiency improvements exists. Analyzing peak power demands also helps identify ways to level those peaks and

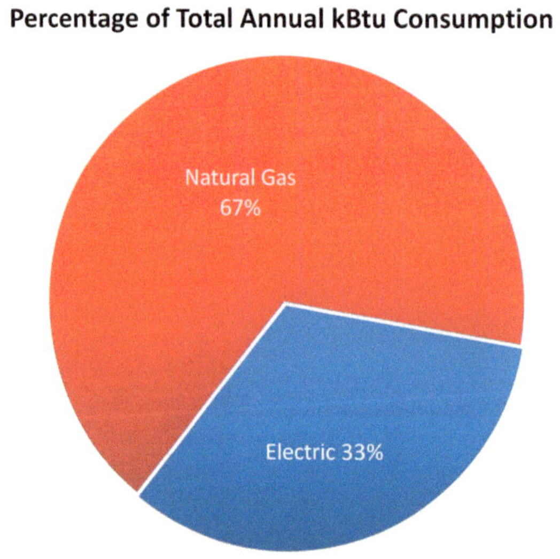

Figure 5 — Example of Annual Consumption for Each Utility Type

potentially downsize HVAC equipment and capitalize on significant financial savings from reduced demand charges.

Energy Costs and Utility Rate Structures

To building owners and operators, reducing building energy costs is equally, if not more, important than reducing energy use. To fully understand the cost of energy consumed by a building, the auditor must determine the rate structure under which that energy use is billed. Energy rate structures vary widely, from extremely simple usage charges to very complex time-of-use structures. Analyzing the current billing rate structure as well as the alternate rate structures that may be available can lead to recommendations before the site visit such as energy demand rescheduling, avoidance of late payment penalties, avoidance of energy ratcheting errors, or switching to a new rate structure.

Building owners and operators do not always understand the various rate structures that control the costs of the energy they consume. The auditor can help the customer to understand the basics of the costs and how to control them successfully. The following areas provide starting points to understanding utility rate structures.

- **Energy Consumption Charges:** Energy consumption charges are typically billed monthly per unit of energy used by the building. Such charges may include customer charges, energy charges, demand charges, power factor charges, and other miscellaneous charges. These charges may vary from month to month.

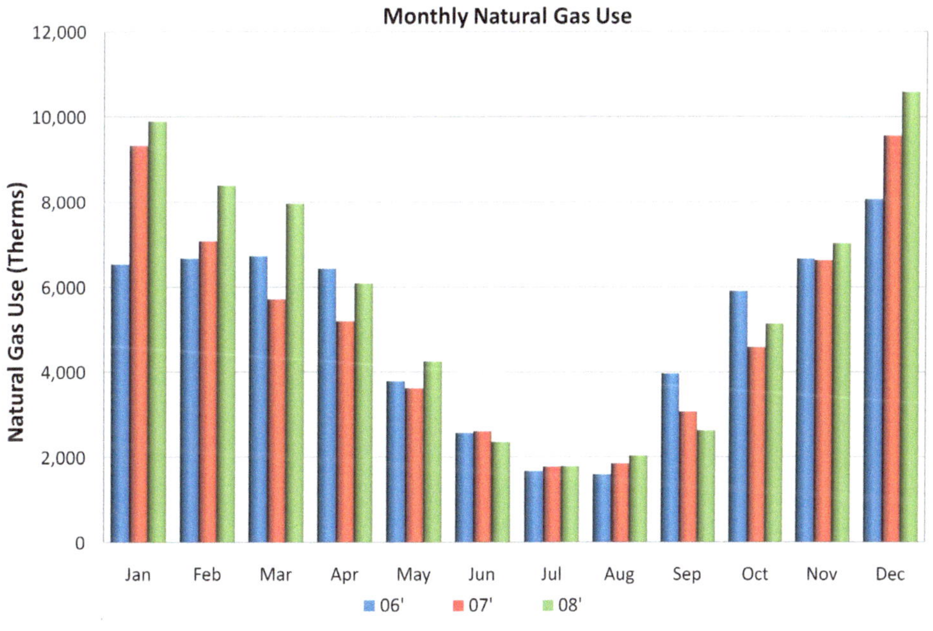

Figure 6 — Example of an Annual Utility Profile

- **Electric Demand Charges:** Demand charges are determined by the maximum power demand in kilowatts that a building requires each month. The demand is typically set during a period varying from 15 minutes to one hour.
- **Ratchet Clauses:** These clauses are used by some utility companies in their rate structures. The clauses allow the utility provider to charge a customer based on the maximum demand or a percentage of the maximum demand from the previous year.
- **Power Factor Charges:** These charges are additional charges for having a low or less-than-optimum power factor, which is the ratio of real power (kilowatt) to apparent power (kilovolt-ampere) for any given load and time. A low power factor affects the utilization of the installed capacity of the electric system. Power factor charges are often structured as additional demand charges or can be per kilovolt-ampere reactive (KVAR) charges.
- **Block Charges:** Block charges are rate structures where a certain unit cost is charged based on different blocks of energy consumption. Energy use within the first block is charged at one rate and then usage above that block is charged at a different rate. Blocks can be either increasing or decreasing. In increasing blocks, the unit price of each succeeding block of usage is higher than the previous block. Decreasing blocks have the opposite cost structure, with the unit price of each succeeding block of usage lower than the previous block. Energy cost savings with utility rates with block charges will largely depend on whether the block is an increasing or a decreasing block, since decreasing blocks do not encourage energy conservation.
- **Time-of-Use Rates:** Time-of-use rate structures are ones in which the unit cost for energy changes during different times, which are typically associated with peak and off-peak periods. Prices can vary based on the time of day, week, season, or year and are higher during peak periods and lower during off-peak periods. Since time-of-use rates are designed to encourage energy conservation during peak periods, peak-shifting strategies used in the energy model can result in significant energy cost savings.
- **Other Charges:** Utilities often charge additional taxes and surcharges based on local regulations and/or programs, such as energy conservation and low-income assistance programs. Additionally, there can also be fuel adjustment charges, which are related to the cost of resource energy to the utility. Often this charge is an additional multiplier that is applied to the energy charge and will vary monthly based on fuel cost fluctuations.

Energy bill costs should be broken down into the components that can be understood and controlled by the owner and operator. These cost components can be listed individually in tables and then plotted into graphs, as shown in

Figure 7. These data are used during the PEA to identify potential measures and will ultimately influence which EEMs are recommended.

Energy End-Use and Cost Allocation

Energy end-use and cost allocation, also called *disaggregation*, is the separation of utility data into general categories based on building energy-using equipment. This exercise identifies categories that will result in the greatest energy-use and cost reductions. This analysis requires utility data and preliminary or estimated building equipment information and operating hours. A whole-building energy model is extremely useful when disaggregating energy into annual end-use categories because it provides reports that automatically summarize this end-use disaggregation.

Building energy end-use categories should be created for both electric and natural gas uses. These categories typically include the following: space cooling, fans, pumps, interior lighting, exterior lighting, plug and process loads, space heating, domestic hot water (DHW), and other specific equipment such as refrigeration systems, kitchen appliances, and swimming pools. Energy use and demand is allocated to each category based on either a whole-building energy model or a simplified method that uses the size of the equipment in the building and its operating hours. With either method, the total energy and costs associated with the building equipment should be within 5% to 10% of the utility data. Figures 8, 9, and 10 provide examples of electric energy and demand allocation using the simplified method.

Utility data disaggregation creates energy and demand baselines for each energy end-use category. Once the end-use breakdowns have been determined, equipment in categories of high energy use can be targeted during site assessments. In addition, the baseline information can be used later when calculating

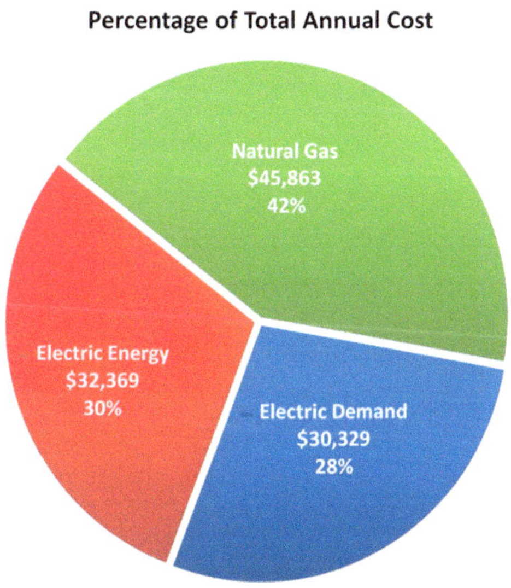

Figure 7 — Example of Total Annual Cost for Each Utility Type

Equipment	Peak kW	% of Total Demand	Monthly Demand Cost ($)	Total kWh	% of Total Energy	Monthly Energy Cost ($)
Fans	37.3	22%	$ 559.65	207,820	23%	$ 7,274
Mech Cooling	70.0	42%	$ 1,050.70	238,000	27%	$ 8,330
Heating	-	0%	$ -	-	0%	$ -
Lighting	47.3	28%	$ 710.61	274,588	31%	$ 9,611
Kitchen	-	0%	$ -	-	0%	$ -
Pumps	12.2	7%	$ 182.50	90,775	10%	$ 3,177
Plug Loads	10.5	6%	$ 157.91	52,603	6%	$ 1,841
Misc	-	0%	$ -	-	0%	$ -
	177.3	**106%**	$ 2,661.38	863,786	**97%**	$ 30,233

Figure 8 — Example End-Use Breakdown

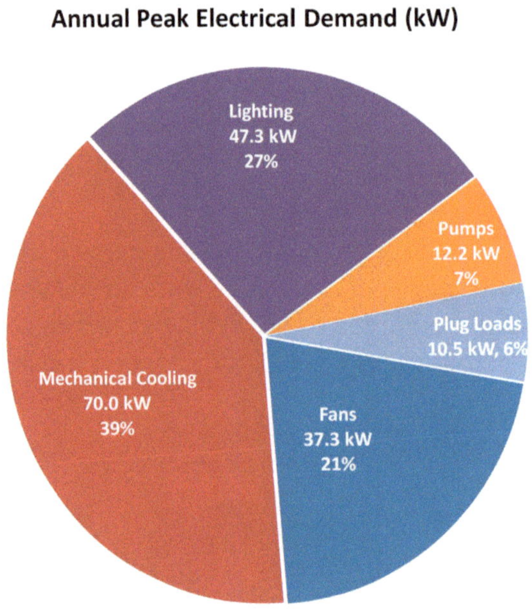

Figure 9 — Example of Electric Demand Allocation

energy savings. The calculated energy reductions should be compared to the baseline allocated energy use to ensure the savings are reasonable.

Benchmarking

The PEA includes creating energy benchmarks for the building. These benchmarks are established based on historical utility data and building characteristics. Benchmarks allow the building to be compared with similar buildings and may assist in determining the benefits of further analysis.

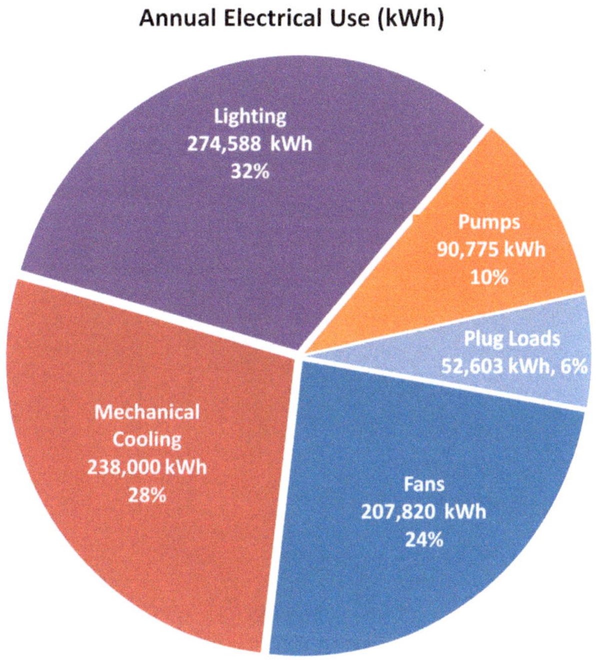

Figure 10 — Example of Electric Energy-Use Allocation

One commonly used benchmark is the Energy Utilization Index (EUI). This number is the energy consumption per unit area of the building. The building EUI can be compared against information available from the Commercial Buildings Energy Consumption Survey (CBECS) maintained by the Energy Information Administration (EIA) (EIA 2011). The CBECS database contains energy use data organized by building usage, climate region, year constructed, and many other factors. The CBECS database also contains typical energy end-use breakdowns for various building types. These should be used as a quality control check against the end-use breakdowns that were determined by the methods described in the Energy End-Use and Cost Allocation section. Use the PEA Template found in Part 3 and at www.ashrae.org/PCBEA to assist with analyzing historical data and comparing key metrics, such as EUI, to various baselines and benchmarks. An example of energy benchmarking results, compared to a distribution of buildings, is shown in Figure 11.

ENERGY STAR® (EPA 2011b) is also used as a popular benchmark for commercial buildings. The Portfolio Manager (EPA 2011a) score upon which the label is based was developed by the U.S. Environmental Protection Agency (EPA) and uses CBECS data on commercial buildings, building characteristics, and weather. A score can be compared to the scores of other, similar buildings to determine the potential for energy reductions. If the building achieves a Portfolio Manager score greater than 75 the facility is eligible for ENERGY STAR certification and a plaque.

For some building types, it may be appropriate to compare a building to its direct peers rather than a sample of similar buildings. For example, for a chain retailer, the energy bills of all the stores in a chain may be a more appropriate

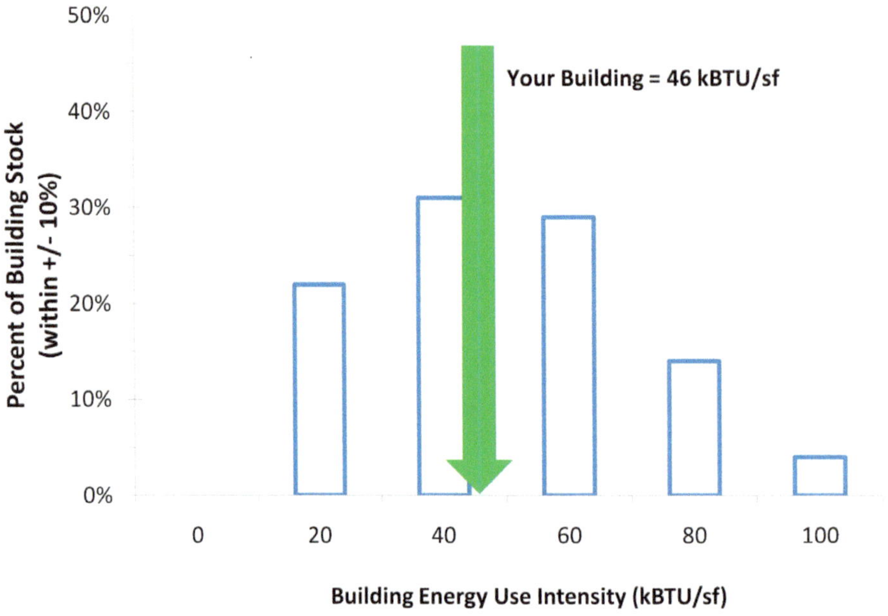

Figure 11 — Sample Building Benchmark

sample for benchmarking than a general collection of retail buildings, because the chain stores will have very similar use, equipment loading, size, and occupancy. If a retailer uses benchmarking in this way it is typical to apply weather normalization if the sites are distributed in dissimilar climates.

Site Visit

Site visits are a critical part of any energy audit. The purposes of a site visit may include

- gaining an understanding of building energy use, occupancy, operation, and occupant behavior;
- reviewing important energy-using systems, processes, and equipment;
- gathering existing equipment nameplate information and controls strategies (use the PCBEA Sample Forms found in Part 3 and at www.ashrae.org/PCBEA to assist in data collection);
- developing preliminary EEMs;
- collecting quantitative data (for example, measuring surface temperatures with an infrared thermometer, as shown in Figure 12); and
- identifying other areas where additional quantitative data are needed.

The following sections provide information for site visits at various stages of the study. It is up to the energy auditor to determine the information required for a particular site depending on the level of investigation and the needs of the individual client.

Figure 12 — Spot Measurement of Hot-Water Piping Temperature to Find Approximate Heat Losses

Preassessment Interview

Prior to the site visit, it is beneficial for the auditing team to begin the data collection process. This will help determine areas of focus during the site visit and help team members understand the energy-use patterns at the site. At a minimum they should follow the procedures in the section titled Preliminary Energy-Use Analysis in Part 2 of this book. If available, project team members should review monitored data for hourly energy consumption to help determine the building's energy use during occupied and unoccupied hours.

The audit team may choose to visit the site before the energy audit is scheduled to take place. This visit may be part of a pre-bid walk-through or as a means of preparations for the site visit. Alternatively, the audit team may choose to conduct a phone interview to collect some of this information.

During the preassessment work, the team may

- discuss the upcoming audit with the site staff,
- give the site staff a list of information that will be needed,
- discuss site access procedures,
- discuss the use of photography during the site investigation,
- conduct a PEA to determine the historical energy use of the building,
- benchmark the building energy use with the energy use of similar building types,
- select the systems in the building to be investigated, and
- collect additional building documentation available (as-built plans, equipment schedules, submittals, prior energy studies, etc.).

Interviews

Interviews with site staff are an essential part of any comprehensive energy audit. The building engineers or technicians that staff the building typically know valuable information about the systems, how they are operated, and the history of changes made at the site. They may have an intimate knowledge of systems that do not operate properly or work-arounds that may be in place. The knowledge that they can contribute to the audit process is very valuable and their input is important to building a team.

In addition to collecting specific information about equipment and controls, the assessment team should ask about the following items during their on-site interviews:

- Equipment condition
- General maintenance procedures
- Persistent comfort issues
- Indoor air quality (IAQ) problems
- Malfunctioning equipment (including excessive O&M expenses, excessive noise, etc.)
- Planned upgrades and capital expenditures for infrastructure
- Potential EEMs
- Previous energy assessments

During these interviews, the building auditor can begin to develop a teamwork approach that will help serve the energy-efficiency retrofit process for the long term. The assessment team should keep this in mind early in the process. Staff at the site may feel threatened by the presence of the assessment team and/or think that the recommendations will reflect poorly on their job performance. The assessment team should stress that they are playing an objective role to determine the best course of action and are not there to evaluate staff performance or place blame. In fact, the results of the assessment may be used to justify equipment upgrades or controls enhancements that will help the staff be more cost-effective and that will provide better long-term maintenance.

The auditor also needs to critically evaluate the information and opinions provided by the site staff. Team building is critical to winning trust from staff who may perceive the audit team as a threat to their job or a threat to the status quo. The auditor should be aware that some questions may put the team on the spot, and every effort should be made to avoid an adversarial approach. If the site staff feel they are being critically evaluated by the audit team or are under the scrutiny of a superior, they may feel pressured to answer questions that are outside their experience rather than answering "I don't know." Building engineers may also be motivated to provide "the right answer" if the audit is perceived as "an interrogation by the energy police." For example, data logging at sites sometimes reveals that equipment is enabled continuously, even after site staff have assured the auditors that runtime controls were enabled. This situation could be due to a misconception on the part of the site staff, controls errors, or a misrepresentation on the part of the staff.

Photographs

Photographs are useful to an energy assessment in many ways. With the advent of digital photography, it is faster and cheaper to record data using a camera than ever before. Collecting information quickly this way helps to make the auditor's time on site much more effective.

Most energy auditors use a camera to record the condition of the equipment on site and to help them remember details of equipment configurations. Sometimes photographs may show details that may not have been noticed on site. They may also help the auditor analyze space requirements or other construction details that may be important potential complications to retrofit measures.

In tightly enclosed or hard-to-reach spaces, often a photograph is the safest and quickest way to gather the needed information. For example, to record nameplate data from a supply fan motor mounted on top of an air-handling unit, it may be easier to reach the nameplate with a camera than to bring a ladder and otherwise access the data.

Photographs are also frequently used to show energy-saving opportunities or equipment in need of maintenance or repair. Common examples are coils that need cleaning, missing fan belts, lights on during the daytime, or bypassed variable-frequency drives (VFDs). Figure 13 shows a photograph documenting the valve position of a building's chilled-water loop. The VFD at the site is pumping against this partially closed valve, which should be open in a variable-flow loop. Inserting photographs of such situations into the audit report helps bring attention to the situations and document the existing conditions.

Photograph best practices include the following:

- Always discuss in advance where photography may be used.
- Never compromise safety when taking photographs.

Figure 13 — Photographs Help Document Existing Conditions and Gain the Attention of Decision Makers

How to Photograph a Nameplate

A digital camera can be the energy auditor's best friend—photographs of nameplates can be a very quick way of recording a lot of information at a site. However, if done incorrectly, the auditor might leave the site with no information rather than a bounty of it. Here are a few hints to help ensure the needed information is collected:

- Photograph the nameplate at an angle or offset from the center to avoid flash "white out" (reflection).
- Use the camera's "macro" setting for close-up views.
- Review the photograph at the site, before leaving, to verify the information is captured.
- Turn the flash off for photographing screen shots of BAS systems
- Rest the camera on something solid to avoid blur when the flash is off.

- Keep a log of photograph locations and times.
- Develop habits that help tie photographs to equipment. For example, to set the context, you may want to take a picture of a piece of equipment from a distance just before taking a picture of the equipment nameplate.
- Keep spare batteries or a spare camera with you.

Building Automation System

A building automation system (BAS), also called an energy management system (EMS) or a building management system (BMS), is the backbone of the control system for larger buildings where centralized controls are used. Most modern BAS systems have trending and data collection capabilities that are very useful for the energy auditor. A screenshot of a BAS user interface is shown in Figure 14.

The BAS makes it possible to quickly collect a lot of information about the energy-using systems in the building. Most automation systems show temperatures of air handlers, zones, the outdoor air, and chilled-water systems, as well as the statuses of pumps, boilers, chillers, and fans. Some systems may also show real-time energy usage of major systems or the current draw of a large motor. Reviewing the BAS with the site staff and discussing the major systems is a good way to learn about the control sequences for the principal systems and find out how the building is run.

The trending capabilities of a BAS are very useful. Most systems are capable of recording and reporting the values of points at defined time intervals or upon changes of state. This capability is a very powerful forensic tool for the auditor. For example, examining the trends for outdoor air temperature, return air temperature, and mixed air temperature can help to diagnose whether an outdoor air economizer is functioning.

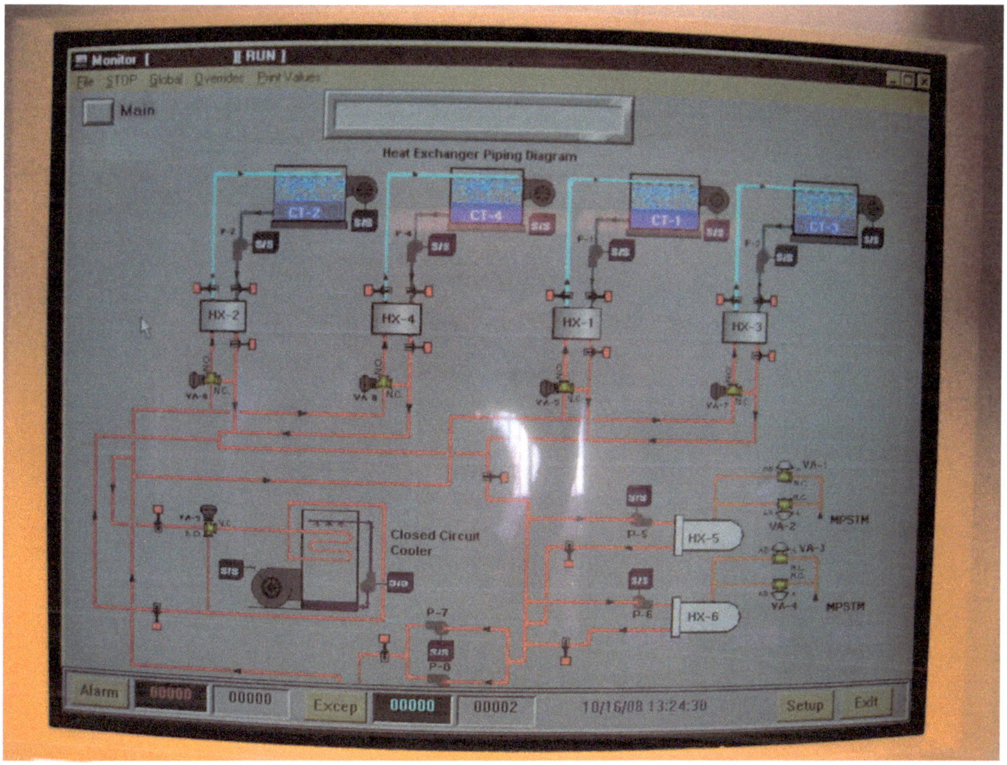

Figure 14 — The Building Automation System's User Interface Frequently Includes Useful Information about System Layout

The energy auditor should also be aware of potential weaknesses involved in relying on BAS data, as a BAS is designed primarily for control rather than for data collection. Sensor calibration can be a limitation, and control output signals may or may not reflect the actual state of the equipment in the field. Output signals are just that—control output signals sent to field devices. Until the effect of the control is confirmed at the control point, only control input states can be relied upon.

An example of trend data collected by the BAS is shown in Figure 15. The frequency of trend data capture should be adjusted to meet the needs of the investigation. Most BASs allow the data collection interval to be set as needed. Setting this interval for a long period may not provide the resolution needed to identify some control issues. However, setting the interval too short may quickly exceed the data storage capabilities of the system and cause excessive network communication that may slow the response of some older systems. For example, to trend outdoor air temperature, a monitoring interval of 15 minutes is typical because the temperature profile is smooth and interpolation in between observations is reliable. However, to troubleshoot the control sequence of a valve that is "hunting," a 15-minute interval may be too long to see the control response; thus, in this case, an interval of 1 minute or shorter may be needed.

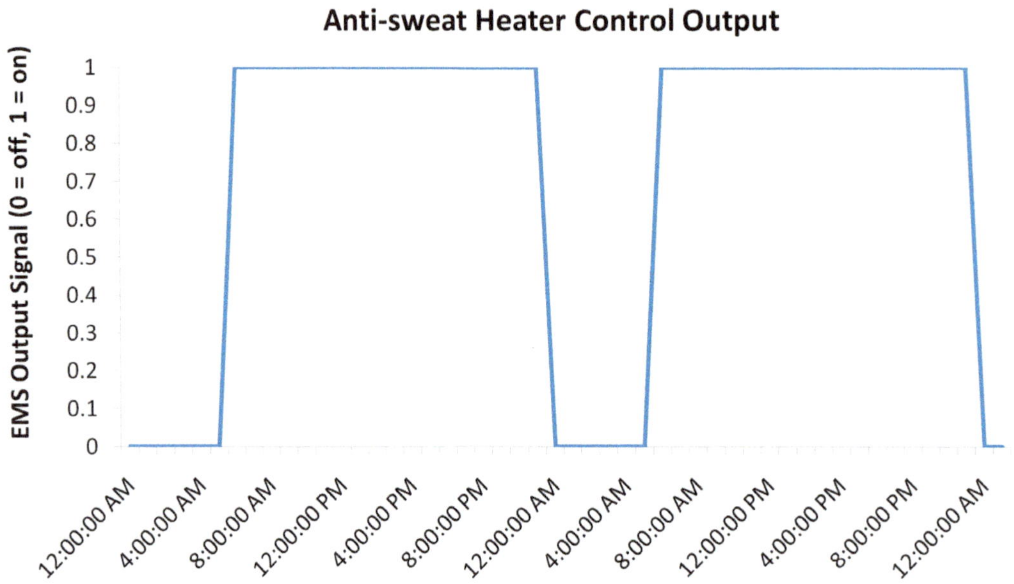

Figure 15 — Sample Trend Data for Antisweat Heaters at a Grocery Store

The BAS is also a convenient tool with which the auditor can review equipment control schedules. The auditor should keep in mind that the schedules may or may not have the intended control results. There may be local switches, disconnects, timers, or bypasses that override preset schedules. A potential pitfall is shown in Figure 16, where a disconnected control wire prevents the EMS from turning the lights off at the contactor as intended. Confirming operation with portable stand-alone loggers can help avoid pitfalls.

In summary, a BAS may be helpful in investigating

- scheduling for mechanical equipment and lighting;
- design control sequences;
- existing setpoints;
- existing control strategies, including supply air temperature control and chilled-water/heating-water/condenser-water temperature control;
- economizer operation;
- simultaneous heating and cooling;
- an overview of the operation of fans and pumps (VFDs, damper and valve controls, variable-air-volume [VAV] systems, etc.); and
- room and/or zone temperature control.

Preparation Activities

Proper preparation before the first site visit will help the team make the most of the time on site. It is important to be efficient at the facility to reduce the audit costs and to minimize the impact on staff resources.

In preparing for a site visit, it may make sense to do a brief phone interview with site staff, as discussed previously. If the staff can make building plans

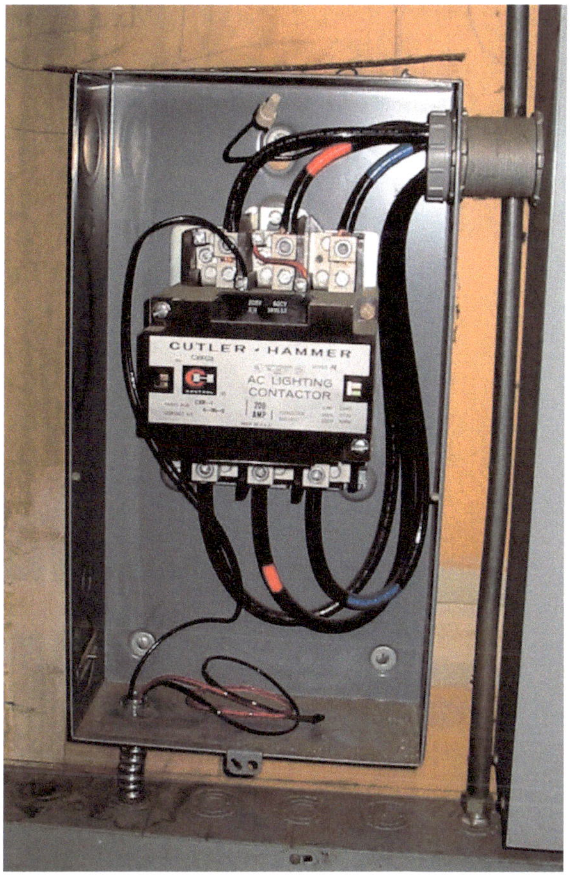

Figure 16 — Can You Trust the BAS Control Output Signal to this Contactor? (Note the Disconnected Control Wire near the Bottom of the Photograph.)

available prior to the audit, this can save a great deal of time reviewing them on site. It may also be advantageous to complete the preliminary energy analysis or benchmarking to have an idea of the approximate energy use intensity of the facility and how it compares to other typical buildings of the same type. Review the PCBEA Sample Forms in Part 3 and at www.ashrae.org/PCBEA and determine what data need to be collected for the specific project.

Typical data that should be assessed prior to a site visit include the following:

- Facility data: building name, use, floor area, year of construction, renovations, and schedules of use (daily and weekly), as well as the number of building occupants
- Site drawings
- Online mapping images
- Utility rates and suppliers
- Utility metering (where energy and water meters are and how metering data are available)

- Energy consumption data (in kilowatt-hours, British thermal units, therms, gallons, or pounds of steam, as appropriate; two to three years of data is preferable)
- Monthly consumption
- Monthly peak demand data (in kilowatts or therms per hour)

Access to some sites may require special security procedures that should be addressed in advance of the initial site visit. Energy audits at correctional facilities (jails and prisons), government buildings, laboratories, military installations, medical facilities, manufacturing plants, museums, and many other building types require that energy auditors prepare in special ways. High-security areas may require background checks, advance drug testing, mandatory safety training sessions, or other special procedures. At a minimum, most sites now require all building visitors to sign a log with security staff at the facility. The auditor should make sure that such concerns are addressed prior to the first site visit to avoid delays.

Online mapping tools now make it possible to inspect buildings using satellite photographs or "bird's eye" views. These images can be very useful for getting a quick overview of a facility's general shape and orientation. In many areas of the United States, free online tools have resolutions sufficient for discerning major rooftop HVAC equipment and looking for opportunities for rooftop solar systems.

The final preparation activity is to gather the equipment and information to take to the site. The exact tools and materials taken to the site will depend on the energy auditor and the level of the evaluation at the site.

For a typical Level 1 Audit, the energy auditor may bring the following:

- Identification
- Notebook/clipboard
- Phone
- Audit forms appropriate for the site
- Digital camera
- Multi-tool
- Flashlight
- Digital thermometer and/or humidity meter (Figure 17)
- Hearing protection (for mechanical rooms)

For a Level 2, Level 3, or more detailed targeted audit, the auditor may also wish to bring the following:

- PCBEA Sample Forms (found in Part 3 and www.ashrae.org/PCBEA)
- Illuminance meter (light meter)
- Infrared temperature spot meter
- Electronic ballast spinner/detector
- Power meter and/or power loggers
- Temperature loggers
- Light loggers

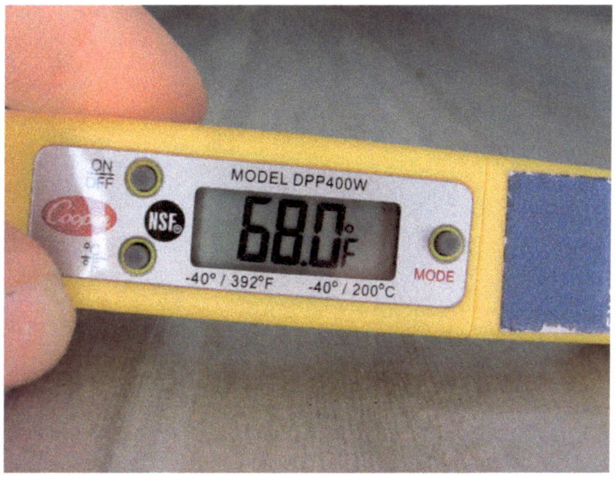

Figure 17 — Spot-Checking Air Temperature with a Handheld Digital Thermometer

- Motor runtime status loggers
- Personal protection equipment (PPE)

Specialized equipment for detailed assessments may also include the following:

- Boiler stack gas analyzer (for combustion tests)
- Ultrasonic flowmeters (for monitoring chilled-water, condenser-water, or heating-water flows)
- Infrared camera
- Pipe caliper
- Centralized data logging systems
- PPE per the Occupational Safety and Health Administration (OSHA); may include construction hard hat, safety glasses, gloves, etc.

Initial Walk-Through

A walk-through audit is typically a brief investigation of a facility that takes the auditor from half a day to a full day to perform. Such walk-throughs can reveal important details (see Figure 18). On-site activities may include those listed in Table 2.

Each of the topics in Table 2 are discussed in more detail in the following subsections.

Detailed Audit Procedures

The procedure for conducting a detailed site investigation (e.g., a Level 2 or Level 3 audit) follows the same basic activities as outlined for a walk-through audit. The primary difference is depth. During a detailed investigation, the auditor goes into much greater depth to collect information to assess the site energy use and to estimate energy savings and project costs for proposed recommendations. Figure 19 shows an energy auditor collecting detailed information on the existing lighting fixtures in a facility. Refer to the PCBEA Sample

Figure 18 — Site Inspections can Reveal Important Details. This Valve was Found Partially Closed just Downstream from a Recent VFD Installation.

Table 2 — Overview of On-Site Activities for an Initial Walk-Through

Initial Interview	Discuss with site staff to review the existing equipment, control practices, maintenance practices, expected projects, previous studies, and priorities.
Site Drawing Review	Collect data from as-built drawings to confirm gross and conditioned floor areas and to determine design intent, equipment capacities, and layout.
Inspection	Collect data on existing equipment and current temperatures and status, inspect equipment conditions, and photograph nameplate information.
Exit Briefing	Review potential EEMs with site staff and discuss their feasibility as well as the next steps.

Figure 19 — An Energy Auditor Conducting a Detailed Inspection of Existing Lighting to Determine the Lamp Type

Forms in Part 3 and at www.ashrae.org/PCBEA for templates to assist with this comprehensive data collection.

The length of on-site investigation for a detailed audit can vary greatly depending on the size of the facility, the depth, and the collection of measured data, such as spot measurements, BAS trends, or stand-alone data loggers. Investigations of small facilities may be only half a day or a day, while detailed Level 3 investigations for large buildings can take weeks.

One of the principal challenges for the energy auditor is to effectively manage time on site. For most clients, the cost-effectiveness of both the energy audit and the recommendation implementations are important. The auditor must find the appropriate balance between time and cost. Greater depth can result in more high-quality recommendations for a facility. However, more time equates to a greater cost—this should be balanced against the additional value provided to the client. The final set of recommendations to the client should be cost-effective when the costs of the audit are included in the total project cost estimates for identified measures.

Table 3 provides a brief overview of the type of information that should be collected on site during a detailed investigation. The table is intended as a guideline, and the auditor should tailor his or her investigation to the potential opportunities and the needs of the client. See the PCBEA Sample Forms in Part 3 and at www.ashrae.org/PCBEA for further details on the data that should be collected on site for different types of energy-using systems.

The energy assessment team may have access to design documents for the buildings they visit. If available, they are very useful to the assessment team. However, for many existing buildings, particularly older structures, they may not be available.

Useful material to be found with the building construction documents may include the following:

- Site plan
- Building-specific layout plans (not necessarily to scale)
- Construction drawings showing building envelope details (materials and construction)
- Mechanical schedules including capacities and design airflow rates
- Control schematics for mechanical systems
- Electrical schedules
- Lighting fixture and lamp schedules
- Lighting control diagrams
- Equipment specifications and manufacturers' submittals
- List of buildings and systems connected to the central BAS (points list)

The information contained in the drawings and supporting information is important for understanding the building's original design, operation, and needs. However, it is important to verify that the equipment and controls shown on the drawings are still present and functioning at the site. Original equipment may have been replaced, which may or may not be documented on the draw-

Table 3 — Summary of Information to Collect On Site

Building Summary Info • Conditioned floor area by space function • Occupancy schedule • Mechanical equipment schedules • Lighting schedules • Utility service • Building use • Summary of systems	**Building Envelope** • Main structure types (drawings, structure layer description) • U-factors of walls, roof, floors • Window and door types, window and door areas by orientation • Condition of sealing/infiltration • Roof condition and type (reflectance)
Central Plants—Cooling • Chiller types/capacities • Chiller sequencing • Chilled-water flow (constant/variable) • Condenser-water flow (constant/variable) • Pumping capacities and sequence • Chiller part-load method	**Central Plants—Heating** • Boiler types/capacities • Boiler sequencing • Heating hot-water flow (constant/variable) • Pumping capacities and sequence • Boiler output control type
Unitary Heating and Cooling • Equipment type/capacity • Control sequence	• Unit schedules • Equipment efficiency
Lighting • Inventory of equipment/type • Record of existing lighting levels • Controls (e.g., photocells, occupancy sensors, manual switching, timers)	**Plug Loads** • Inventory of computers/servers/monitors • Office equipment type and density • Vending machines
Air Handling • Inventory of equipment/type • Supply airflow • Heating/cooling capacity • Ventilation/exhaust rates • Zone controls • Ventilation rates in typical spaces	**Domestic Water** • System type (boiler/water heater) • Storage capacity • Recovery rate • Recirculation type • Pumping capacities and sequence
Controls • BAS • Trending capability • Controlled points/equipment/zones • Pneumatic versus direct digital control (DDC) equipment	**Other/Miscellaneous** • Electric distribution schematic • Uninterruptible power supply (UPS)/ transformers • Backup power • Cogeneration on site • Thermal storage • Renewable generation • Vertical transport (elevators, escalators) • Kitchen equipment

ings at the site. Control strategies are frequently changed by building operators. It is important to investigate existing control sequences during the interview with building staff, by logging equipment operation, and/or by reviewing control sequences through the BAS.

Practical Arrangements for Site Work

The energy auditor should make sure that the site staff is aware of the practical arrangements for the site visit well before the visit occurs. Good communication before the visit and ensuring that the team has access to the site can save a lot of time during the actual field work.

Before the team arrives, the site staff should be aware of the following:

- Who is in charge of getting the information the team needs before the visit
- Who will be present when the team is on site
- Where the team will work (working space and facilities)
- Who attends in-briefings and exit briefings
- Whether the team will have free access to the buildings or need an escort
- Restrictions for operating on site
- Whether the team will have their own keys to technical rooms
- How the building's occupants will be informed of the assessment team's visit

The energy auditor may also need to check with site security to determine what site-specific requirements there are for the auditors to have access to the site. Some questions to consider include the following:

- Are identification or passport details or copies needed in advance?
- Are background checks required?
- Is there any special training (e.g., on site-specific safety) required?
- Are equipment lists with equipment identification details needed for bringing measuring devices to site?

Making these arrangements prior to arrival at the site may save critical time during the investigation.

Reviewing Initial Recommendations/Exit Briefing

It is very useful to conduct an exit briefing at the site to review the initial recommendations and collect final information from site staff. This is usually performed as a final step but may be performed earlier to accommodate schedules for site staff who are familiar with existing equipment and controls.

The exit briefing includes a presentation of preliminary energy conservation measures (ECMs) that the audit team has noted during the investigation. This discussion typically does not include any analysis or cost estimates except typical values from similar facilities.

The primary aim of the briefing is to review recommendations and solicit feedback from site staff pertaining to the feasibility of the proposed recommendations. This discussion helps foster teamwork with the site staff, which is essential to successful retrofit projects. Discussing potential EEMs can also provide vital information to the audit team while laying a foundation of understanding with the site staff, who will be responsible for maintaining energy savings. There may also be "next steps" that the site staff can implement immediately if simple controls or behavior changes are identified in the audit.

As a final step, the audit team manager should also briefly discuss the project schedule, outlining when final results and reports will be available. Following the briefing, the site staff should have an idea of what to expect next and any immediate steps that might be taken to improve energy efficiency.

Safety

Conducting site visits in commercial buildings involves risks. The practice of inspecting energy-using systems may take the auditor into environments where there is open mechanical equipment, unsafe electrical wiring, or a number of unforeseen conditions. When on site, the auditor should observe safe operating practices appropriate for the environment he or she is in.

While the following list is not exhaustive, it does include a few practices that are particularly appropriate for energy auditors practicing in commercial buildings:

- Don't do anything that is beyond your level of experience, qualifications, or training.
- Don't enter any environment that you feel is unsafe. For example, avoid exposed wiring, unventilated enclosures, and poorly lit areas.
- Always know the location of the nearest exit.
- Beware of spinning machinery with exposed shafts or fans.
- Be careful on entering pressurized mechanical rooms or walk-in plenums requiring considerable force to open doors, as sudden and forceful closing is difficult to control and fingers can get caught in door jambs upon closing.
- Always use PPE appropriate for the environment, as demonstrated in Figure 20). This includes ear plugs or similar hearing protection for compressor rooms, eye protection around open electrical panels, and gloves rated for installing current transducers.
- Wear clothing that is suitable for the working environment.
- Heed warning signs, equipment lock-outs, and alarms.
- Most refrigerants are heavier than air—beware of unventilated areas around chillers and refrigeration systems. Check refrigerant alarm statuses.
- Ammonia is caustic; avoid exposure to the eyes, skin, and respiratory system.

Figure 20 — Use Appropriate PPE when Installing Data Logging Equipment

Photograph courtesy of FPL Energy Services, Inc.

Measurement

Data are vital to quantifying operating parameters and performance. The first requirement is to define the information needed to characterize the systems and support further analyses. Data might range from setpoints and conditions (e.g., temperatures and humidity) to loads (e.g., heating load, outdoor air ventilation load, etc.) to efficiencies (e.g., boiler efficiency and cooling coefficient of performance [COP]). Types of data might vary from one-time snapshots to load profiles and periodic totals. Defining building characteristics and the level of required accuracy are key to defining data needs.

Total building energy use is a starting point to characterizing building performance and is often available through energy bills or metering systems. The next level segregates energy use into main end-use categories such as lighting, heating, cooling, fans and pumps, water heating, and miscellaneous. This usually requires submetering to quantify, although loads such as lighting can be estimated based on connected load and scheduled operation. On a system level, energy use and operating conditions can be quantified through measurements either specifically for an energy audit or through existing logs and/or control system data. At the finest level, equipment operation and performance can be discerned through measurements to determine frequency of use, load levels, and efficiency. The ASHRAE publication *Performance Measurement Protocols for Commercial Buildings* (ASHRAE 2010a) provides recommended measurement protocols at three levels of detail for energy, water, and indoor environmental quality (IEQ).

The following sections describe types of data that are frequently required to identify and analyze efficiency measures.

Operational Characteristics

Setpoints

Setpoints can be determined from the building controls. However, some values might not be accessible to the user, requiring someone with access to the controls programming. Setpoints should be reviewed for consistency with how the building is intended to be operated. Typical setpoints include the following:

- Space temperature
- Space humidity
- Space lighting levels
- Minimum outdoor airflow rates
- Carbon dioxide levels
- Economizer limits
- Boiler temperature
- Chiller temperature
- DHW storage and delivery temperatures
- Fan system flow control static pressure
- Water loop system flow control static pressures
- Heating and cooling system enable conditions

Operating Schedules

Operating schedules are readily available through the building controls. However, some parameters might not be user adjustable and are only accessible through a review of the controls logic. A frequent opportunity for energy savings is turning equipment and systems off when not needed. A review of the actual operating schedule compared to the intended operation of the building is vital to a thorough audit. Other operating schedules may offer opportunities for energy savings. Their settings or the omission of the settings should be noted. Examples of schedule-related data are as follows:

- Occupied/unoccupied hours in each controlled zone
- Warm-up and cooldown periods
- Unoccupied override conditions and timers
- Flow control resets (e.g., fan static pressure with VAV box dampers)
- Equipment operating temperature resets (e.g., boiler with outdoor air)
- Cooling tower condenser water resets

Operating Conditions

Loads

The load conditions under which equipment operates can impact efficiency based on how the equipment deals with part-load operation. Furthermore, the total loading directly impacts the energy use through the system efficiency. The timing of loads can also be critical in setting peak electric demands, impacting energy pricing and cost. Thermal loads can present difficulties in that they require the combination of multiple measurements, usually flow and temperature differences. The flow measurement in itself might need to be inferred (e.g., from pressure drops, etc.) rather than directly measured. Some typical loads of interest to energy efficiency are:

- Heating and cooling loads
- Peak loads
- System start-up loads
- Lighting power density
- Fan and pumping power loads

Equipment Performance

Equipment efficiency directly relates the loading to the energy use. Attention must be paid to the impact of part load on efficiency. For example, the efficiency of a condensing boiler system is highly dependent on the return water temperature of the system. Any field-measured efficiency levels must document the operating parameters (e.g., temperatures, loads, etc.) under which the efficiency measurement is taken. Efficiency adds another level of complexity to the load measurements in many cases, requiring the auditor to determine energy input simultaneously with a measured loads. Typical efficiencies to be evaluated are

- combustion efficiency,
- cooling efficiency, and
- energy recovery efficiency.

Some of this information involves simple one-time readings of gauges, control system sensors, or handheld meters, while others are calculated values dependent on combinations of measurements (e.g., load from water flow and temperature difference) and still others involve a trend of data from continuous monitoring.

Data are available through many sources with varying degrees of usefulness. Some buildings have extensive data available through control systems (for example, Figure 21 shows the efficiencies of three chillers in a building under various cooling load requirements). Other buildings might only have unitary controls or thermostats with no recorded data capabilities. The auditor must assess the information needed to complete the analysis as well as the available data and decide on the value of collecting additional data.

Spot Measurements

A snapshot of data is most useful to represent conditions with little variation, such as space temperature and light levels (see Figure 22). The connected load of a lighting circuit with all the lights on or airflow in a constant-volume fan system, for example, are represented well by a single sample measurement. Typical operating values might be discernable through trend data from a building control system.

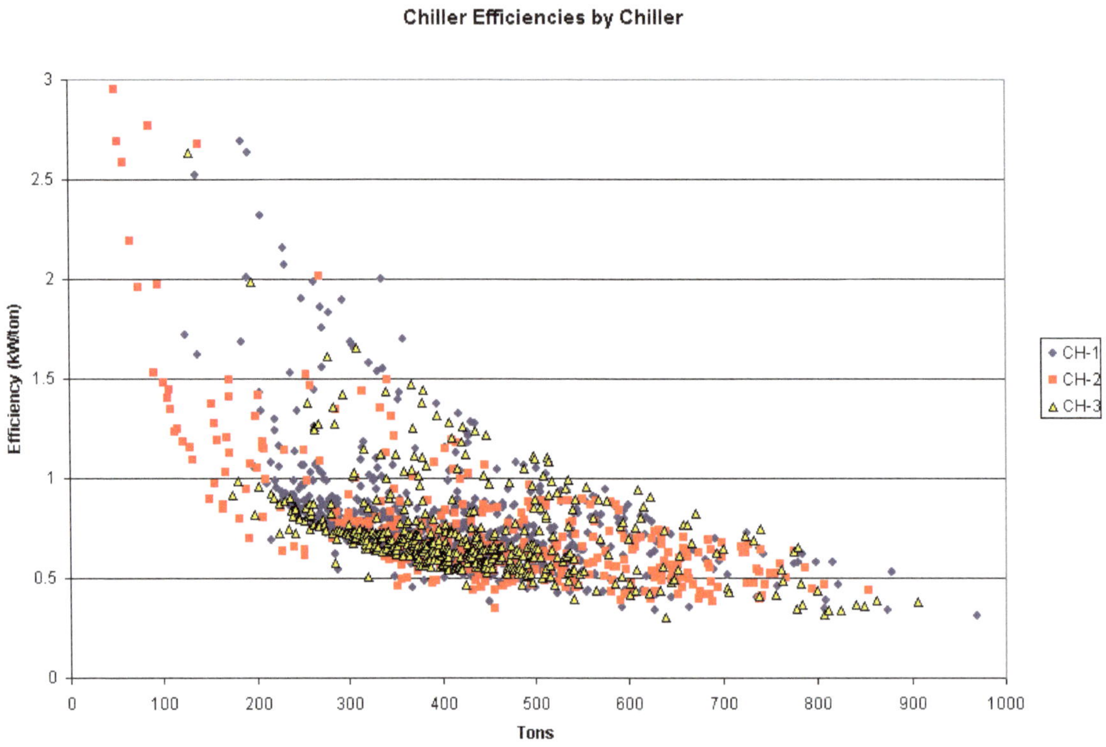

Figure 21 — Time Series Data Can Show Revealing Trends

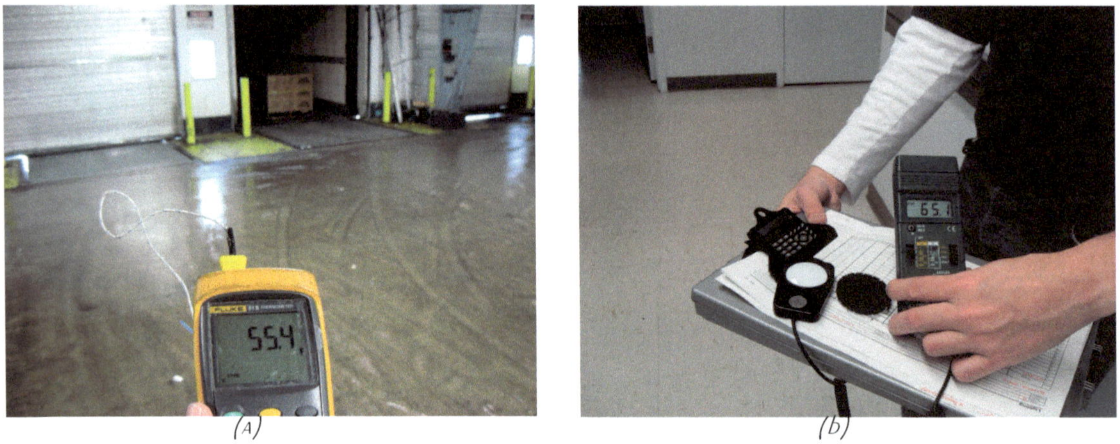

Figure 22 — Example Spot Measurements: (a) Space Temperature and (b) Light Levels

Interval Measurements

Trend data are useful in verifying operations within schedules or relationships between driving factors and system responses (e.g., outdoor air temperature reset of heating hot water). Measurements may be taken over a short or a long time frame.

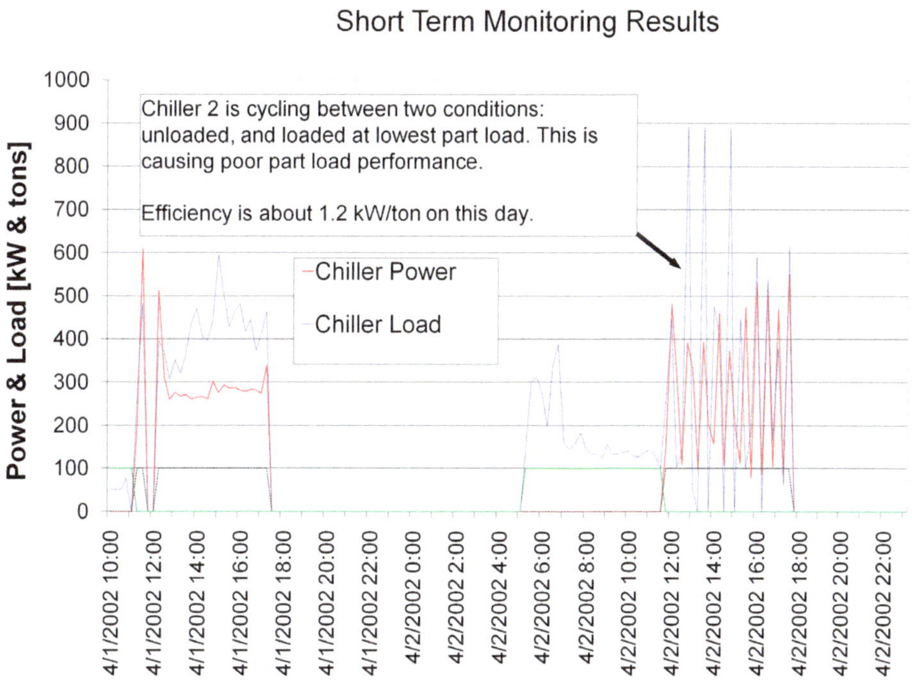

Figure 23 — Example Short-Term Interval Monitoring

Duration

Long time-frame (e.g., monthly) data provide insights into seasonal impacts on energy use and system operations. Short time-frame data might cover a few hours to a day or a week to verify the operating schedule and reactions to conditions. A weekly trend of lighting power may help confirm the impact of daylighting control if, for example, a sunny day might be captured over a week's duration. A daily trend would confirm that systems start and stop in alignment with the building schedule.

Interval

Measurement intervals of 15, 30, and 60 minutes are common in building control systems (see Figure 23). These interval lengths give a good indication of how the building follows operating schedules and some indication of how the systems respond to changes in operating conditions. Longer intervals provide a trade-off with less data storage at the expense of providing less detail. A shorter interval (e.g., 1 minute) may be justified for a short duration in order to assess the dynamics of control system tuning.

Building control systems often offer change-of-value or change-of-state trends. These variables record only the date and time when a piece of equipment cycled on or off or a value changed by a predetermined amount. These types of trends are very compact (i.e., the stored size of the data is small) and offer precise records to assess cycling of equipment, but due to the inconsistent time interval they can be difficult to deal with if there is a desire to compare them to other trends.

Annual, monthly, and daily intervals are useful in looking at total energy use. They provide a high-level view of operations useful for benchmarking and comparisons to historical data. Daily data are particularly useful in developing weather correlations, where the daily total electricity or gas use is correlated to the daily average outdoor air temperature. To some extent monthly data can be weather normalized using heating and cooling degree-days to determine the portion of energy use that is weather dependent. Annual totals provide a basis for historical comparison.

Instrumentation Options

Opportunities to easily collect data vary widely with each building. The following sections discuss potential data sources.

Building Control Systems

Most building control systems offer some level of data trending. Data may be archived in a server for years for every point or may be limited to the last 24 hours of a few predetermined points. The ability to quickly assess the availability and scope of data is a good skill for an auditor to develop. The building operator or the controls specialist are key sources of information. Their levels of familiarity with the building control system features will provide the basis for determining data and might provide insight into the level of effort in ongoing evaluations of systems operations through the control system.

Some systems might be underutilized with few defined trends. Therefore, it might be worth a small investment in time to learn about setting up trends on the particular control system with the building operator.

Some systems store overwhelming amounts of data. Knowing the key operating parameters for the main systems being reviewed is vital in not having to deal with large amounts of data. Most systems can present a listing or graphic representation of data; however, it is often useful to capture the data for further off-line analysis in a spreadsheet to develop relationships among multiple points. Most systems offer data export features to common spreadsheet or text formats. Data in varying time frames and formats can be aggregated and analyzed by publicly available tools such as the Universal Translator (UTOnline.org 2009).

Graphical representations of the systems can be useful, but it is important to verify that the graphics are representative of the actual system. Many systems have standard graphics that generally apply to a particular site but sometimes omit specific details. The graphics can provide an easy means to capture snapshots of an entire system's operation. Often screen captures or hard-copy printouts are the only means to capture these data.

Stand-Alone Data Loggers

Data loggers are available as single-measurement battery-operated devices or as multimeasurement configurable devices. When no trend data are available from a building control system, these stand-alone loggers (see Figure 24) serve well to quantify operations. Runtime, status, temperature, humidity, and electric current are the most common stand-alone data loggers, but loggers exist for

Figure 24 — Installing Stand-Alone Power Logging Equipment

Photographs courtesy of FPL Energy Services, Inc.

capturing data from any sensor with an output signal. Data intervals are often user configurable and duration is limited only by the amount of onboard memory and the precision of the data. Data loggers offer an advantage to building assessors in that their calibration can be maintained, whereas when trend data are used the calibration must be established anew for each building site.

Handheld Meters

Handheld instruments with a known calibration condition provide both a good verification means of existing data and a source for new data. Temperature, humidity, carbon dioxide, airflow, lighting level, sound, amperage, and power are all readily measurable with handheld equipment. For example, lighting fixture power can be measured by a handheld meter, as shown in Figure 25. Water flow rate can be measured by a calibrated bag and timer, as shown in Figure 26. Accessibility and the level of intrusiveness often limit the ability to use handheld meters.

Existing Gauges

Like building control system sensors, existing gauges provide the same data used by the building operators. The accuracy of the gauges both in measurement and in proper placement to be representative are vital (see Figure 27). Using faulty data to assess operation will not provide any new insights to the systems. The auditor should consider the need to verify gauges through comparison with calibrated handheld equipment and verification of sensor placement based on the importance of the data (see Figure 28). For instance,

Figure 25 — Measuring Lighting Fixture Power Using a Handheld Meter

Photograph courtesy of FPL Energy Services, Inc.

Figure 26 — Measuring Water Flow Rate Using a Simple Calibrated Bag and Timer

Photograph courtesy of FPL Energy Services, Inc.

Figure 27 — Correct Installation of an Ultrasonic Flowmeter is Required for Meaningful Data

Photograph courtesy of FPL Energy Services, Inc.

Figure 28 — An Auditor Verifies Chiller Operating Conditions on a Built-In Monitor

understanding whether the hot-water temperature gauge is representative of the boiler temperature or the delivery temperature in a system that has both primary and secondary loops is important to clarify in documenting system operation.

Accuracy

Erroneous data will lead to erroneous results. Accuracy must be considered for the individual measurement as well as the application of the measurement to be representative of the intended information. An independent comparison to a calibrated measuring device provides a high level of verification. Sometimes it is possible to make relative comparisons of sensors measuring the same values when systems have no loads or energy inputs (e.g., supply and return water temperatures across a chilled-water system that has chillers disabled).

One should assess the needed level of accuracy. For instance, a discharge air sensor from a VAV box need not be very accurate, as its primary function is to verify the operation of reheat (i.e., that the delivered air warms when heat is called for and does not warm when heat is not called for). On the other hand, the accuracy of a building static pressure sensor used to control an exhaust fan speed would be vital to the building pressurization and energy use. Many sensors used in control systems function in closed control loops, so the absolute accuracy is not as vital as the ability to be repeatable and stable. For example, in a system where supply duct pressure is maintained by varying the fan speed, the absolute value of the pressure has little significance if the duct pressure setpoint is reset based on air terminal damper positions. However, if the duct pressure is maintained at a fixed absolute value, a sensor might drive the system to operate at a needlessly higher pressure, using more fan energy.

The cost of monitoring increases with the increasing complexity in data collection (Figure 29). It may be beneficial to review calculations and findings

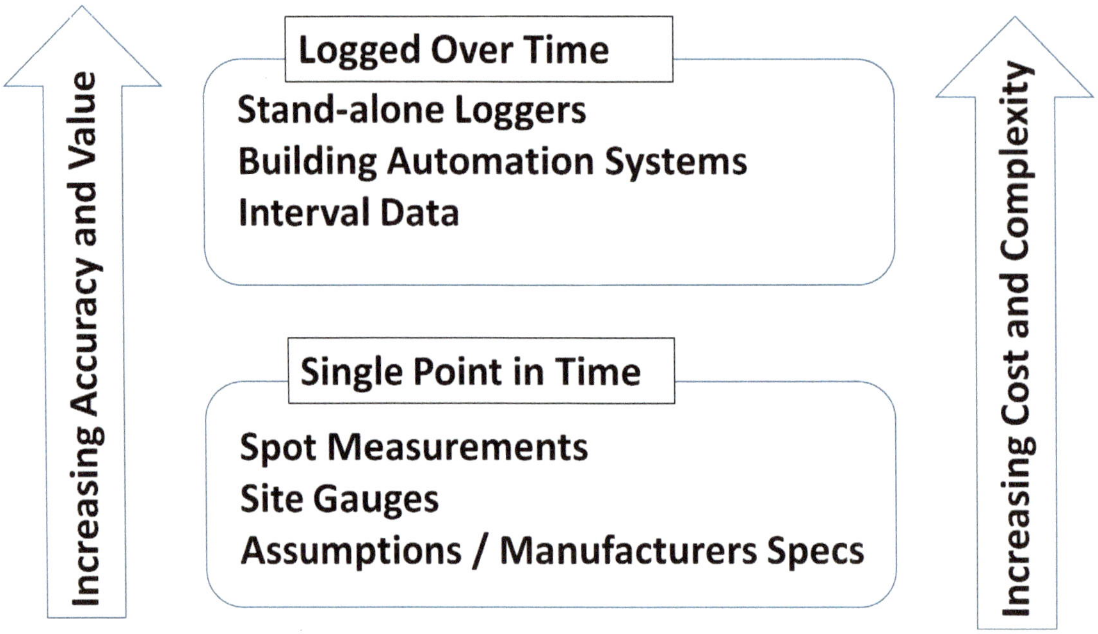

Figure 29 — Cost and Complexity for Measurement Strategies

Measurement Resources

International Performance Measurement and Verification Protocol (IPMVP) provides an overview of current best-practice techniques available for verifying results of energy efficiency, water efficiency, and renewable energy projects (EERE 2002).

M&V Guidelines: Measurement and Verification for Federal Energy Projects provides guidelines and methods for measuring and verifying energy, water, and cost savings associated with federal energy savings performance contracts (FEMP 2008).

ASHRAE Guideline 14, *Measurement of Energy and Demand Savings*, provides guidelines for reliably measuring the energy and demand savings due to building energy management projects (ASHRAE 2002).

ASHRAE Handbook—Fundamentals, Chapter 36, "Measurement and Instruments," describes characteristics and uses of instruments used to measure building performance (ASHRAE 2009).

ASHRAE Standard 105, *Standard Methods of Measuring, Expressing, and Comparing Building Energy Performance*, is intended to foster a commonality in reporting energy performance of existing or proposed buildings to facilitate comparison, design, and operation improvements and development of building energy performance standards (ASHRAE 2007a). This standard is currently under revision. Users should refer to the most recent edition for up-to-date guidance for key data that affect results to determine what the impact of data uncertainty has on conclusions. Then, the effort to verify accuracy can be focused on only key parameters that influence the survey conclusions.

Analysis

Overview

Determining the most appropriate calculation methodology and energy analysis tool is perhaps one of the most challenging and important steps in the energy audit process. The method must be capable of evaluating all design options with sufficient accuracy to make correct choices. *ASHRAE Handbook—Fundamentals* (ASHRAE 2009) states that the following factors should be considered when choosing a calculation methodology or energy analysis tool:

- Accuracy
- Sensitivity
- Versatility
- Speed and cost
- Reproducibility
- Ease of use

The auditor should begin outlining the calculation methodology and selecting the analysis tools based on the level of the energy audit and its requirements. A Level 1 walk-though type of energy assessment does not involve whole-building energy simulation or assess detailed measures from large quantities of data. Likewise, the potential energy savings for this type of audit is based on site observations, engineering assumptions, brief calculations, research, and experience. As the audit involves more comprehensive investigations, the savings calculations can become more detailed. Thus, energy savings identified at Level 1 are understood to be potential or possible energy savings to be assessed at Levels 2 and 3. Energy modeling or simulation typically is not done until Level 3. With Level 2 and 3 audits, the data collected and organized in the PCBEA Sample Forms (found in Part 3 and at www.ashrae.org/PCBEA) can serve as the inputs for the energy and financial analyses.

Energy Efficiency Measure Types

An energy efficiency measure (EEM) is a collection of actions and equipment installations intended to reduce a building's on-site energy use while simultaneously maintaining or improving IEQ, service, and operating requirements. Recommended EEMs will result in economic benefits, primarily through reduced energy costs. They may also result in O&M cost savings. Additional nonenergy benefits such as improved reliability or occupant comfort may also occur.

EEMs contrast with ECMs, which may reduce energy consumption by lowering service levels. An example is de-lamping. A de-lamping action used if

Efficiency vs Conservation: Are They the Same?

In energy auditing practice, *efficiency* and *conservation* have slightly different meanings. *Efficiency* refers to providing the same level of service with equipment that uses less energy. *Conservation* can apply to measures that offer a lower level of service or comfort. However, the terms are often used interchangeably in practice.

EEMs reduce energy use while maintaining or improving IEQ. For example, a new boiler that has a higher combustion efficiency will provide the same (or better) performance with the same heat output but with lower input energy. ECMs save energy by turning equipment off or down. If a measure continues to satisfy the building mission and meet acceptable service levels, an ECM is also an EEM. For example, if lights currently are on 24 hours a day but will be controlled so that they are only lit during occupied hours, this conservation measure also improves overall efficiency without decreasing service levels for occupants. ECMs that reduce service below minimum requirements—for example, measures that lower light levels to those less than needed for occupant tasks—will save energy but will sacrifice environmental quality. Such draconian recommendations must be avoided.

An auditor should ensure that any proposed measures, EEMs or ECMs, will not compromise occupant comfort or safety or reduce service levels below limits set by codes and accepted building practice.

the space was initially overlit might be a viable energy efficiency option if Illuminating Engineering Society of North America (IES) light levels were maintained. However, energy conservation actions such as turning off computers, coffee pots, etc., to reflect occupancy patterns would not really lower service levels.

No-Cost/Low-Cost Measures

No-cost/low-cost EEMs are EEMs that can be implemented within the O&M budget for a facility. They frequently involve modifying how a building is operated. The modifications can be as simple as scheduling equipment off when not needed; a more complex example is the optimization of a building HVAC control strategy that involves a coordinated change to setpoints, schedules, flow rates, and other system variables. Just because an efficiency measure is low cost to implement does not imply that the savings that result from the change are simple to calculate.

Simple no-cost/low-cost measures may be implemented by building managers in conjunction with standard O&M activities, while more complex projects may require controls specialists or energy engineers as part of a full or partial existing building Cx process. No-cost/low-cost measures require the use of trained building operators or energy engineers but do not involve significant capital costs or the replacement of large pieces of equipment.

Capital Investment Measures

Capital investment EEMs involve the replacement of energy-using equipment or the reconfiguration of existing systems. Examples include removing existing chillers and installing more efficient models or converting a constant-volume air-handling system to a variable-volume system. Implementing the measures can involve extensive demolition and construction and require the use of HVAC, electrical, plumbing, controls, and other contractors. Capital investment projects involve significant costs, which are recouped through reduced energy costs. Construction of capital investment measures may require careful planning to minimize interruptions to building operations.

Demand Response Measures

Demand response measures result in temporary reductions in energy use as requested by a distribution utility during a peak demand period. Customers enter into an agreement with the utility to curtail energy use when requested and in exchange are paid for the resulting demand savings. The reduction is short term and results in a lower demand only. Demand reduction strategies may have associated energy savings, but the strategy can sometimes increase overall consumption in the building. The short-term demand savings may also decrease service levels in the building, such as through higher space temperature setpoints or decreased lighting levels.

Demand response measures are not EEMs, though they can be cost-saving opportunities for building operators. Some examples of demand response measures include increasing temperature setpoints during hot weather or temporarily reducing light levels. Demand response opportunities only exist when customers can continue to function at below-normal power requirements for short but unpredictable periods and where the distribution utility offers to pay for the reduction. Customers may incur a penalty for failure to curtail upon request. If an opportunity is implemented every day in the building, it is an energy efficiency measure, not a demand response measure. Demand response, by definition, is a strategy that is only intended to be implemented a few days of the year to avoid peak energy consumption during periods when generation supplied to the power grid is near its maximum and costs are at their highest.

Demand Shifting and Load Shifting Measures

Demand shifting and load shifting measures use equipment and control strategies to shift energy use from peak demand periods to times when energy costs are lower. Demand shifting and load shifting measures are similar to demand response measures in that they manage a building's power requirements in response to a distribution utility's economic signals. Like demand response measures, they also usually result in little energy savings and may actually increase energy use. Customers implement demand shifting/load shifting measures to take advantage of differential pricing schedules where energy costs rise and fall in response to system demand. The price schedules can be fixed, as in a time-of-use rate schedule where customers pay more during weekday nonholiday peak hours, or they can be dynamic, as in a real-time-pricing rate schedule where customer costs reflect changes in wholesale costs that fluctuate hourly with system demand.

Examples of demand shifting/load shifting measures include temporary, voluntary changes such as shifting production schedules to evenings and permanent, capital-intensive strategies such as using thermal storage. Demand shifting and load shifting measures are viable only when differential energy prices are offered; therefore, their analysis should include an assessment of the probability of rate schedule change.

Self-Generation Measures

Self-generation measures use on-site generators to improve operational reliability or to offset the purchase of utility-supplied electric energy. Self-generation equipment is powered by burning either fossil fuels or renewable fuels to drive an engine, turbine, fuel cell, or other device connected to an electric generator or by photovoltaic panels that convert solar energy to electric current. Self-generation measures may require significant investments, detailed analysis of energy rates and operating costs, and compliance with environmental regulatory requirements. In commercial buildings, self-generation is most commonly installed as standby or emergency generation to meet reliability needs; it may also allow operators to peak shave by generating electricity on site rather than paying peak rates during a period of high system demand. Typical self-generation efficiencies range from 15%–30% and rarely exceed those of utility-supplied power. Self-generation may provide an economic benefit if electric rates vary with system demand. For these reasons it is typically dispatched on a short-term basis and not operated to meet base load energy needs.

Cogeneration Measures

Cogeneration measures combine on-site electrical generation with the recovery of the thermal energy released by burning fossil fuels or renewable fuels or through the operation of fuel cells. The recovered thermal energy that would otherwise be exhausted to the atmosphere is then used to offset fuel requirements for thermal loads. Like self-generation, cogeneration is usually expensive, involves specialized maintenance, and must comply with environmental and other permitting or operating regulations. Determining a return on investment requires a careful analysis of the costs of electric and thermal energy supplies and a building's electric and thermal energy loads. Unlike self-generation equipment, cogeneration equipment operates on a continuous or near-continuous basis, supplying base load requirements for both electric and thermal energy. In commercial buildings, cogeneration capacity is usually constrained by the usability of the recovered thermal energy; unused thermal energy quickly drives down the efficiency and cost-effectiveness of the investment. Cogeneration is common in hospitals or laundries with large, continuous thermal loads; it may also be cost-effective in large buildings with absorption chiller systems.

EEM, ECM, ECO, EEO: What do They Mean?

Energy engineers use a number of closely related terms to describe the opportunities, equipment, and actions to improve a building's energy efficiency. These include energy efficiency measure (EEM), energy conservation measure (ECM), energy conservation opportunity (ECO), energy cost reduction measure (ECRM), energy cost saving measure (ECSM), and energy efficiency opportunity (EEO). The terms are often used interchangeably but can have slightly different connotations. ASHRAE has selected *energy efficiency measure* as the phrase most descriptive of the actions and equipment by which building energy efficiency is improved. *Conservation* can imply savings achieved by reducing services or operations, e.g., turning off lights during unoccupied hours, and can have a connotation of doing less with less. *Opportunity* logically describes only the potential for savings, though in practice it is often substituted for *measure* and all the activities associated with installing equipment or optimizing schedules.

Energy Calculation Methodology

The energy calculation methodology outlines the principles and procedures that are used to determine energy savings and economic impacts. A methodology for measuring energy is essential to ensure a consistent and uniform approach. The main parties involved in the energy audit should agree on this process. Typically this will consist of the client and the energy assessment team but may also include other program managers such as representatives from utility companies.

Although the techniques may vary from audit to audit, key elements will be considered for the calculation methodology in every energy audit. These should be carefully considered before selecting the analysis tools. The elements, listed below, are described in greater detail in the following sections:

- Energy end-use analysis
- Utility rate treatment
- Measure interactions
- Measure order
- Measure lifetime
- Cost estimates
- Utility incentives
- Tax incentives
- Weather data normalization and annualization of results
- Analysis tools

Energy End-Use Analysis

An end-use analysis (Figure 30) helps the energy auditor identify and account for all building energy inputs and outputs. End-use analyses may be calculated (with whole-building annual energy simulation tools or spread-

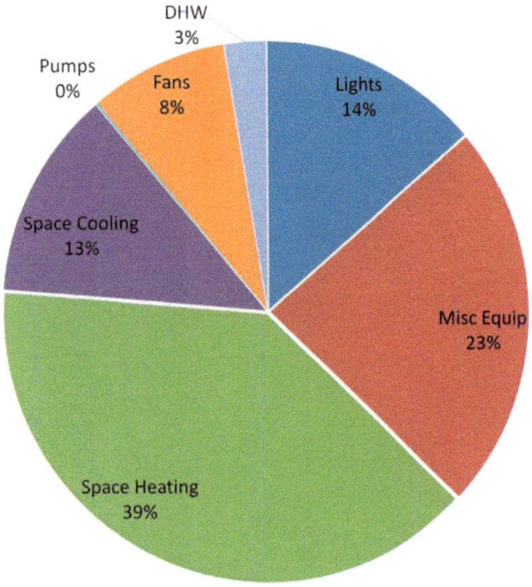

End Use	kWh	therms	kBTU	%
Lights	85,668		292,300	14%
Misc Equip	145,457		496,300	23%
Space Heating		8,336	833,600	39%
Space Cooling	78,605		268,200	13%
Pumps	645		2,200	0%
Fans	52,550		179,300	8%
DHW		547	54,700	3%
Total Estimated	**362,925**	**8,883**	**2,126,600**	**100%**
Historical Billing	**366,455**	**9,217**	**2,172,044**	
Percent of Actual	**99%**	**96%**	**98%**	
Total per sq ft	**11.9**	**0.3**	**69.5**	

Figure 30 — Example End-Use Analysis

sheets) or measured. The calculations are based on assumptions for climate, usage, and system operation. Measurements capture the actual operating conditions and account for factors such as equipment inefficiencies and operational errors.

An energy balance is a related process for establishing the reasonableness of energy consumption estimates. In an energy balance, the analyst compares his or her estimate of the facility's energy use with the historical bills for the facility. There should be a reasonable agreement between the estimates of the base-case energy use and the billing data. If there is not, then the analyst's assumptions are incorrect or there are pieces missing from the building's energy use puzzle. For conservative energy savings estimates, the analyst should always use assumptions that lead to an underestimation of the site's base-case energy use rather than an overestimation.

The end-use analysis should include heating and cooling energy consumption as well as energy consumed by lighting, motors, and other plug/process loads.

Utility Rate Treatment

The utility rates used in energy calculations have a significant effect on the cost savings that are associated with each recommended measure. It is essential that energy analysts are well versed in the various components of complex utility rate structures and the impacts of these parameters within the analysis. The utility rate structures should be analyzed during the preliminary energy analysis, before the site assessment is conducted. Beyond just energy consumption charges, all of the following should be considered:

- Fixed monthly charges
- Demand charges
- Power factor charges

- Block charges or tiered rate structures
- Time-of-use rates and charges
- Other taxes and surcharges based on local regulations and/or programs, such as energy conservation and low-income assistance programs

The client should agree to the utility rates that are used to calculate cost savings. In some situations the client may want to use estimated future rates to attempt to capture future savings due to inflation or expected tariffs.

The assessor should plan ahead to use calculation methods that will provide the data inputs needed by the utility rate structure.

Measure Interactions

ECMs usually apply to one piece, group, or type of equipment, while the energy used in a commercial building is dynamic and is based on the interactions of all equipment, internal loads, and external loads. Complex and often unexpected interactions can occur between the systems or various modes of heat transfer. For many measures, savings for one system or energy type may increase or decrease the consumption of another. Before selecting an analysis tool, it should be determined whether interactions will occur, how significant they will be, and how they will be accounted for.

Interactions can sometimes be difficult to quantify in simple analysis calculations. These calculations usually isolate equipment components and may not take into account the entire system operation. If interactions are expected to cause a significant energy penalty or benefit, a more detailed analysis tool (such as an hourly simulation tool) may be most appropriate.

Some of the common interactions are listed below:

- Supply air temperature increase causes fan energy increase
- Chilled-water temperature increase causes pumping energy increase
- Lighting energy reduction causes heating energy increase

A recommended approach to assessing interactive measures is to begin with measures that affect loads then proceed to measures affecting air-side systems and then the central cooling or heating plant, followed by heat rejection. This order of measure analysis allows reductions in successive systems to accumulate so that gains can be realized due to appropriate equipment sizing during each step.

EEMs can also have interactions related to reducing loads on the building or to reducing incremental implementation costs. Considering a number of load reduction measures together can have significant simplifications for equipment downsizing, or even elimination, and subsequent capital cost reductions. For this reason, measures should be evaluated as integrated packages when these effects are significant and not simply be eliminated from an analysis performed on isolated measures.

Measure Order

When a mutually exclusive approach is not used to calculate savings, the order of the measures can have a significant effect on the savings results. The analysis method called *measure stacking* includes changes from previous measures when calculating subsequent measures in an effort to avoid double-counting savings. If measures are not appropriately ordered, or stacked, savings will still be double counted. For instance, scheduling measures are included first so savings of subsequent measures are not counted when the equipment is recommended to be off.

Measure Lifetime

It is important to analyze the remaining service lives of heating, ventilation, air-conditioning, process, and electrical systems. When systems are approaching the end of their useful lives, it should be determined whether implementing a measure is still feasible before the equipment is replaced.

The lifetime of technology associated with building systems should also be evaluated. Changing to current technology provides better solutions for building automation, control, and metering.

Several life-cycle cost analysis (LCCA) tools are available in varying complexity to assist in the analysis of this important factor. The Building Life Cycle Cost (BLCC) Program was developed by the National Institute of Standards and Technology (NIST) to support life cycle cost evaluations (NIST 2010). This program allows a user to compare life-cycle costs for various energy and water conservation measures. NIST Handbook 135, *Life-Cycle Costing Manual for the Federal Energy Management Program* (NIST 1995), also provides a framework for assessors to calculate life-cycle cost in custom spreadsheet applications. This approach may be advantageous when intermediate results are needed or where the calculation methodology must be documented or modified. Other resources include Chapter 36 of *ASHRAE Handbook—HVAC Applications* (ASHRAE 2007b) and life-cycle costing handbooks (e.g., Kirk and Dell'Isola 1995).

Cost Estimates

Accurately estimating implementation costs can be time consuming and generally requires years of experience. These estimates directly affect the economic analysis that clients use to determine which measures to investigate further or implement and to set budgets for those measures to be implemented. The accuracy of cost estimates depends on the audit level and the customer requirements. Discussions with the client prior to energy and cost saving analyses should outline the acceptable level of estimate and the assumptions that are used.

The method for determining implementation costs can vary widely from vendor bids to rule-of-thumb estimates. Several tools are available to assist in estimating implementation costs, such as the commercially available RSMeans data (RSMeans 2011). However, it should be noted that the estimating tools are generally only as accurate as the user's knowledge of cost estimating.

Regardless of the method used to estimate implementation costs, the implementation costs of the capital improvement baseline should be taken into consideration. The capital improvement baseline allows for differentiating the costs of maintaining basic building functionality or meeting other goals not associated with energy efficiency versus significantly improving efficiency. EEMs may be framed in terms of incremental capital cost when a project will be completed for reasons other than energy savings. The assessor may present an energy-saving upgrade over the conventional replacement as an EEM.

Utility Incentives

Utility incentives and rebates are common as utility energy efficiency programs have grown and energy savings goals have increased. Incentives can provide important funding for the building owner to conduct energy studies and implement efficiency measures. These incentives can be distributed during various phases of the energy audit, from completion of the audit report to verification of measure implementation, or even upon confirmation of actual energy reductions.

Utility incentives are primarily effective through a reduction in the implementation cost of an EEM, making it more economically appealing to the building owner. It should be determined whether these savings will be included in the cost savings analysis or in cost estimating. The client may require that the actual implementation cost be shown separate from the potential utility incentive.

Savings calculations that are the basis for utility incentives may differ from the savings calculations used to estimate the customer's economic benefit of a measure. Many utilities offer customized incentives that are based on the energy savings of an EEM and pay a rebate based on measured or calculated savings. The energy savings estimates that form the basis of these incentives are typically calculated using a code-compliant baseline rather than the customer's existing equipment. For example, a utility pays $0.09 per annual saved kilowatt-hour for large commercial energy upgrades. If a customer is replacing a chiller at the end of its useful lifetime, the customer would be required by building code to replace that chiller with a new chiller that meets a minimum performance standard (following ASHRAE/IES Standard 90.1 [ASHRAE 2010], for example). In this case, the energy savings from the utility perspective is the difference between the proposed new chiller and the baseline code-compliant chiller. However, the energy savings that accrue to the building owner are based on the difference between the old chiller and the new chiller. In situations like this, the energy analyst may need to provide two sets of calculations: one to show the owner's expected bill savings and a second to provide the energy savings to the utility for the customized incentive.

Tax Incentives

Tax deductions are currently available (as of the date of publication) to private businesses that install permanent EEMs in buildings when those measures exceed the requirements of ASHRAE/IES Standard 90.1-2001 (ASHRE 2001).

To claim the deductions, the energy savings must be calculated using hourly simulation tools approved by the Internal Revenue Service (IRS). The deductions under the *Energy Policy Act of 2005*, Section 179D (GPO 2005), range from \$1.80 to \$0.60 per square foot (\$19.38 to \$6.46 per square metre) depending on the magnitude of the energy savings, with the maximum deduction being available to buildings that have an energy savings of 50% or more relative to the Standard 90.1 baseline. These tax incentives may change after 2011, so analysts should consult current IRS guidelines before claiming the deductions.

Weather Data Normalization and Annualization of Results

Weather data can be used to adjust estimated annual energy savings. Energy savings are typically calculated based on historical weather that may not accurately reflect the weather for a specific year. Many locations experience unusually warm or cool seasons. During these seasons, the energy use may also deviate from expected results. Energy savings that are not realized can be a significant issue if expected cost savings are used to finance the implementation of measures or utility incentives are tied to measured energy reductions. In these cases it may be necessary to adjust the expected savings based on the actual annual weather data.

Analysis Tools

Today there are a wide variety of energy analysis tools available that range from basic hand calculations to whole-building computer simulations. Not only do these tools have varying levels of accuracy, but also they are intended to be used for different levels of energy analysis and require very different levels of effort and cost. The U.S. Department of Energy (DOE) Office of Energy Efficiency and Renewable Energy (EERE) maintains a directory of building energy software tools on its Web site (EERE 2011c).

It is important that the auditor understands how to select the correct analysis tools with varying levels of information. The accuracy of the savings calculations relates more directly to the detail and correctness of the inputs used than the detail of the tool. The value added by using a more comprehensive calculation tool decreases as more input assumptions are made. Higher-quality inputs are more likely to benefit from use of a more complex calculation methodology.

Steady-State Calculations

Simple and accurate steady-state calculations can be used where assumptions remain constant over time—for example, if the indoor temperature and internal gains are relatively constant and if the heating or cooling systems are to operate for a complete season. The concepts of balance-point temperature and degree-days remain valuable tools.

The balance-point temperature of a building is defined as the value of the outdoor temperature at which, for the specified value of the interior temperature, the total heat loss is equal to the heat gain from the sun, occupants, lights, etc. For many calculations the balance-point temperature represents an average.

The severity of a climate can be characterized concisely in terms of degree-days. Degree-day methods are the simplest methods for energy analysis and are appropriate if the building use and the efficiency of the HVAC equipment are constant. Annual tabulations of degree-days for various climates are published by National Oceanic and Atmospheric Administration (NOAA), ASHRAE, and other public organizations. Historically, an indoor reference point of 65°F (18.3°C) has been used to account for the fact that even the most poorly constructed building is capable of maintaining comfort conditions without heating when the temperature is at least 65°F (18.3°C). Modern commercial and institutional properties may often have a balance point less than 65°F (18.3°C).

Bin Method

The bin calculation method evaluates equipment operation based on outdoor air temperature intervals and the number of annual hours within each interval. For this method, the annual energy consumption is calculated at several outdoor air temperatures and then multiplied by the number of hours.

The weather data necessary for this analysis is widely available for locations in the United States and in growing locations around the world. Annual hourly weather data is usually separated into weather bins of 2°F or 5°F (approximately 1°C to 2°C) increments for calculations.

For many calculations, this method provides accurate savings results compared to the level of information collected during the audit. Performing an hourly calculation or using an energy model instead of a weather bin model will not always result in more accurate savings. Generally, audits are not detailed enough to understand the hourly variations in building loads or equipment operation. Using a more detailed tool will not result in more accurate results if they are based on estimated operation.

The bin method may require refinement in establishing the bins to be used when considering the effects of time-of-use energy rates on building operating costs. Also, bins constructed based solely on dry-bulb temperature can oversimplify the relationship of temperature to humidity for buildings with high quantities of outdoor air or dependence on evaporation as part of the operation of the HVAC system.

Modeling/Simulation

A whole-building energy simulation can capture the dynamic nature of a commercial building. The Quick Energy Simulation Tool, eQUEST (JJH 2009b), is a common whole-building simulation tool used to analyze a building's energy use. Examples of building models created using eQUEST are shown in Figures 31 and 32.

The selection of a building energy analysis program depends on its application, the number of times it will be used, the experience of the user, and the hardware available to run it. The following categories should be considered before selecting a modeling program:

- Complexity of input
- Quality of output required

- Availability of weather data
- Capability of program to model specific features and technologies, such as variable-refrigerant volume systems or thermal mass

The complexity of an hourly simulation model may be required for projects where a deeper level of analysis is required. Examples of scenarios that are best addressed through modeling include the following:

- Building envelope measures
- Comprehensive projects, with a high degree of interaction among measures
- Documenting tax incentives
- Projects where time dependence is important, such as effects of building massing on cooling loads

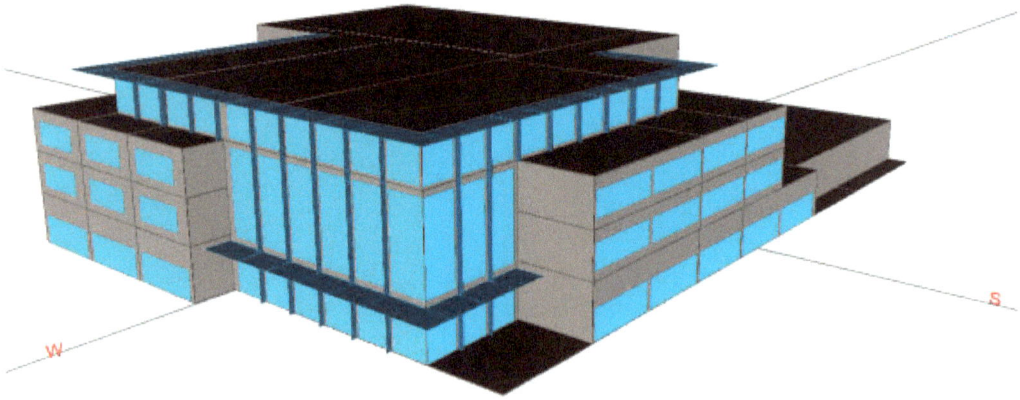

Figure 31 — Visualization with eQUEST, an Hourly Simulation Software (JJH 2009b)

Figure 32 — eQUEST model and Actual Photograph of University of Hawaii C-MORE Lab Building

- Sophisticated rate treatment is needed as with hourly or time-of-use rate schedules
- Incentive program requirements
- New constructions or additions where building loads are unknown
- Documentation of savings to third parties

The following guiding principles should be followed when performing building energy calculations:

- Be knowledgeable of the inner workings of the analytical tools.
- Be knowledgeable of the technologies being modeled.
- Prioritize efforts.
- Follow analytical procedures that facilitate quality assurance.

The results of a computer program should be reviewed carefully for accuracy. The software vendor does not accept responsibility for the correctness of calculations or use of the program. Energy and cost savings should be checked based on the auditor's past experience and knowledge of expected savings. See the Hourly Simulation Checklist in Part 3 and at www.ashrae.org/PCBEA for guidance.

Model Quality Control

Quality control is important in modeling to ensure the results are valid and meaningful. The modeler can provide quality control by

- documenting assumptions and input values,
- using preprocessing tools/spreadsheets to convert component descriptions into modeling input values, and
- making design changes incrementally in the model (run and check often!).

Checking the model output will help identify incorrect assumptions in the base case and verify that the simulated equipment is represented as intended. Some steps for checking model output include the following:

- Develop a review checklist (see topics to include in the below table).
- Review summary output reports.
- Extract data for evaluating reasonableness of results:
  - key output values and
  - metrics, back-of-the-envelope calculations, and hourly data.
- Compare results from output files and report side by side by evaluating
  - rule-of-thumb metrics and
  - performance of actual buildings—billing data or monitored performance.
- Determine whether the changes are as expected.

See the Hourly Simulation Checklist in Part 3 and at www.ashrae.org/PCBEA to help ensure your model is reasonable and accurate.

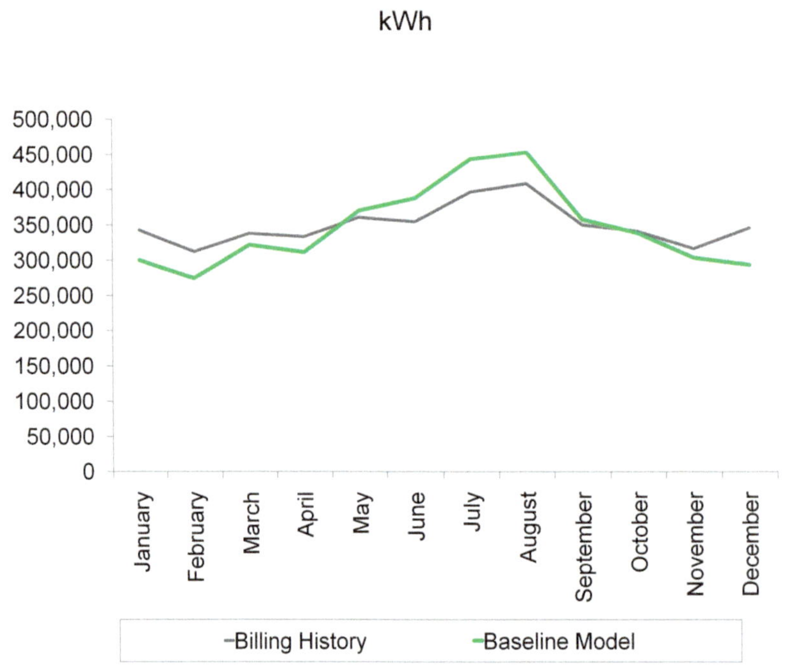

Figure 33 — Sample Calibration to Billing Data

Whole-building energy simulation can be advantageous when many interactions will be present among the assessment recommendations and when adequate information is available to construct the model. The main disadvantage is the time needed to build the model, in many cases requiring information that may not have been otherwise required. The assessor may also encounter difficulty if the whole-building energy simulation must be calibrated to utility bills, particularly in projects with varying uses and occupancies. Figure 33 shows a sample calibration of energy use between the results of the simulation model and the billing history.

Economic Evaluation

There are two basic financial models that analysts can use to make economic evaluations: simple payback and life-cycle cost analysis (LCCA). Both models can evaluate individual measures as well as packages of measures. While a simple payback model is a quick and easy evaluation typically used for very short payback or simple scenarios, a LCCA model accounts for the time value of money and enables the client to compare alternatives with different lifetimes and cash flows and make a much more informed decision. For this reason, the analyst is encouraged to use LCCA modeling to evaluate all major investment decisions. This section provides an overview of the differences between simple payback and LCCA models as well as basic information on the metrics produced by each.

Model Calibration Resources

New construction and major renovation projects can utilize energy modeling to reconcile actual performance with predicted performance and to calculate verified savings. A key step in the modeling process is calibration to actual consumption, which includes reviewing utility billing data, collecting on-site survey data, aggregating building performance data, and fine-tuning assumptions and inputs. The depth to which calibration takes place (whole-building, end-use, system, or equipment) is dependent on the needs of the project and the availability and level of performance data collected. In general, more data translates to more analysis time and greater cost. The cost should be justified based on the value of the data being collected.

The table below outlines supporting resources to help practitioners when calibrating energy models to actual performance data.

Methods	ASHRAE Guideline 14, *Measurement of Energy and Demand Savings* (ASHRAE 2002) ASHRAE Research Project 1051 Final Report, "Procedures for Reconciling Computer-Calculated Results with Measured Energy Data" (Reddy and Maor 2006) ASHRAE Research Project 1404, Measuring, Modeling, Analysis, and Reporting Protocols for Short-Term M&V of Whole Building Energy (ASHRAE n.d.) *International Performance Measurement and Verification Protocol* (IPMVP) (EERE 2002) *M&V Guidelines: Measurement and Verification for Federal Energy Projects* (FEMP 2008)
Tools	ASHRAE Research Project 1050 Final Report, "Development of a Toolkit for Calculating Linear, Change-Point Linear and Multiple-Linear Inverse Building Energy Analysis Models" (Kissock et al. 2002) Universal Translator (UTOnline.org 2009) Energy Charting and Metrics (ECAM) Tool (CCC 2008) Building Optimization Analysis (BOA) Tool (CCC 2010) ASHRAE Research Project 1093 Final Report, "Compilation of Diversity Factors and Schedules for Energy and Cooling Load Calculations" (Abushakra et al. 2001) Building Energy Software Tools Directory (EERE 2011c)
Data Sources	Commercial Buildings Energy Consumption Survey (EIA 2003) EnergyIQ (LBNL 2011) ASHRAE/IES Standard 90.1 (ASHRAE 2010b) energy end-use schedules Title 24 energy end-use schedules (CEC 2008) Published reports and papers providing end-use loads and load shape data Historical weather data Design rules of thumb Performance metrics

Clients will often have their own financial personnel involved in the detailed analysis needed for a major investment decision. In these situations, information will need to be provided to enable clients to perform their own economic analyses.

A simple payback analysis considers only the initial capital cost and annual energy cost savings of a measure. The simple payback method ignores all costs and savings occurring after the point at which the payback is reached and does not differentiate between alternatives with different lifetimes or different up-front capital investments.

In sharp contrast, LCCA is defined by NIST as "an economic method of project evaluation in which all costs arising from owning, operating, maintaining, and ultimately disposing of a project are considered to be potentially important to that decision" (NIST 1995, p. 1-1). An LCCA model for building retrofits can account for the cost of borrowing money, inflation, future replacement or maintenance costs, energy cost escalation, and depreciation as well as their effects on taxes and salvage values. It can also incorporate any planned capital improvements (see the Establishing a Capital Improvement Baseline section that follows) and calculate the marginal cost of the measure or package in relation to this baseline. As a result, the LCCA can take numerous data about the measures, packages, and existing building to generate a highly informed economic evaluation of the proposed investments.

In addition to the scope of analysis, the metrics produced by the two models differ greatly. An LCCA model generates financial metrics that normalize for project lifetime and cash flows over time, such as life-cycle cost (LCC), net present value (NPV), and internal rate of return (IRR). The methods for calculating these metrics have been documented at length in Chapter 36 of *ASHRAE Handbook—HVAC Applications* (ASHRAE 2007b). A brief summary of metrics and their definitions is provided in Table 4.

Establishing a Capital Improvement Baseline

To establish a capital improvement baseline is to understand what building systems or components need to be replaced for reasons unrelated to saving energy over the next several years. There are two basic approaches: 1) the analyst can interview the building owner or facility manager or 2) if there are no planned replacements, the analyst can estimate when replacements are due by using various published sources and his or her own professional judgment of condition. Chapter 36 of *ASHRAE Handbook—HVAC Applications* (ASHRAE 2007b) provides service life estimates for mechanical systems. Life-cycle costing handbooks (e.g., Kirk and Dell'Isola 1995) provide service life estimates for envelope, lighting, and other systems.

There are two principal reasons to establish a capital improvement baseline. First, it informs the financial performance evaluation of measures and packages. The baseline enables the analyst to differentiate for the client the costs of maintaining basic building functionality or meeting other goals not associated with energy efficiency versus significantly improving efficiency. For instance, if the analyst or facility manager expects a boiler to reach the end of its life in about

Table 4 — Economic Metrics and Definitions

Metric	Definition
Time Value of Money	The time-dependent value of money, reflecting the opportunity cost of capital to the investor during a given time period.
Discounting	A technique for converting cash flows occurring over time to time-equivalent values, at a common point in time, adjusting for the time value of money.
Inflation Rate (i)	The average rate at which the purchasing power of a unit of currency decreases each year.
Real Discount Rate (d_{real})	The rate at which future costs are discounted to the present (accounts for the time value of money).
Nominal Discount Rate ($d_{nominal}$)	The rate at which future costs are discounted to the present (accounts for the time value of money and inflation).
Simple Payback Period (SPP)	Length of time needed to pay back the initial capital investment, usually expressed in years. Not a sufficient metric for long-term economic performance.
Simple Return on Investment (SROI)	The inverse of the simple payback period.
Internal Rate of Return (IRR)	The annual yield from a project, usually expressed as a percentage of the total amount invested; the compound rate of interest which, when used to discount cash flows, will result in zero net savings. If the IRR is greater than the investor's stated discount rate, the measure is considered beneficial. Useful for comparing profitability of investments.
Adjusted Internal Rate of Return (AIRR)	Annual yield from a project, taking into account reinvestment of interim returns.
Net Present Value (NPV)	The value (gain minus cost) of an investment in today's dollars over some specified time period. If the investment has a positive NPV, it is generally considered to be beneficial.
Levelized Cost	The present value of the total cost of an investment, converted to equal annual payments.
Savings to Investment Ratio (SIR)	The ratio of the present-value savings to the present-value costs. If the SIR is greater than 1, then the alternative is considered to be cost-effective.
Life-Cycle Cost (LCC)	The total discounted cost of an investment over some specified time period. It includes initial capital cost as well as the present value of maintenance costs, energy costs, and replacement costs, plus the salvage value.

five years, the cost of a premium-efficiency condensing boiler is incremental to the cost of the future replacement (which, to maintain basic building functionality, would be a standard-efficiency unit).

The second reason to establish a capital cost baseline is to indicate to the analyst particular opportunities for efficiency measures. Many planned improvements can be easily redesigned to save a lot of energy and downsize (or even eliminate) mechanical equipment, such as glazing and roof replacements and interior remodels. Similarly, a planned major HVAC equipment replacement or expansion indicates an opportunity to save capital costs by implementing load-reduction measures.

Reporting Uncertainties

It is sometimes advantageous for the auditor to show, for both savings and cost values, the range of reasonably possible values, the basis of the uncertainty, the method of managing the uncertainty, and why selected values were reported. Reporting uncertainties in this way may help secure financing for the project. Energy audit cost and savings estimates should be seen as estimates, not as defined values. Any entity providing financing for a project needs to know the risks in their investments, which may not be identified in the audit report. But the financier may put another layer of conservatism on the project if he or she does not understand the uncertainty in the reported costs and savings. This may have the effect of making the proposed measures look riskier or less cost-effective than intended by the auditor. Since energy auditors are addressing risk managers, auditors need to report results that financiers can understand. They need to report uncertainty clearly and avoid layering multiple safety factors. Any conservatism added by the engineer should be clearly reported to the reader.

Spreadsheet-based tools for conducting LCCA and are available for download online from a number of organizations, such as NIST and Rocky Mountain Institute (RMI). Table 5 provides a list of key economic formulas often used in performing LCCA.

Table 5 — Key Economic Formulas

Metric	Formula
$d_{nominal}$	$d_{nominal} = (1 + d_{rsal}) \times (1 + i) - 1$
SPP	$\text{SPP (years)} = \dfrac{\text{Initial capital cost}}{\text{Annual cost savings}}$
SROI	$\text{SROI} = \dfrac{\text{Annual cost savings}}{\text{Initial capital cost}}$
NPV	$\text{NPV} = \sum_{t=1}^{n} \dfrac{CF_t}{(1+d)^t} - CF_o$ where CF = cash flow, t = current time period, and n = number of cash flows
SIR	$\text{SIR} = \dfrac{\text{NPV}_{savings}}{\text{NPV}_{costs}}$
IRR	$0 = \left(\sum_{t=1}^{n}\right) \dfrac{CF_t}{(1+\text{IRR})^t} - CF_o$

LCCA Example

Consider the January 2011 replacement of an old 500 ton centrifugal chiller that runs for the equivalent of 2000 full-load hours per year:

Category	Value
Time frame	10 years
Discount rate	8%
Electricity rate	\$0.12/kWh
Demand charge	\$10/kW/mo (for 8 months per year)

The key to establishing the baseline in this example is to estimate the remaining life of the chiller and account for the capital expense required to replace it at the end of its useful life.

Suppose the chiller is estimated to have a remaining life of five years, after which it will no longer function. In 2016, the baseline will need to include the cost of a replacement chiller, which we will assume to be ASHRAE/IES Standard 90.1 compliant (ASHRAE 2010b).

Now that the baseline has been established, the measure must be defined. Table 6 lists the critical information to be considered in the economic analysis.

We will also assume that replacing the old chiller with an efficient chiller will save \$5000/yr in maintenance for the first five years. The savings end at year six because at that point the old chiller would have been replaced.

The LCCA yields the annual discounted cash flows, shown in Table 7. As can be seen, the annual savings are large for the first five years and reduce significantly after the installation of the code-compliant chiller. Also, the discounted cost of this chiller can be clearly seen in 2016. Summing up the discounted cash flows results in a NPV of \$34,501 (an ROI of 11.2%).

Table 6 — Chiller Data

Category	Baseline Existing Chiller	Baseline Replacement Chiller	New Efficient Chiller
Efficiency	0.65 kW/ton (COP = 5.41)	0.577 kW/ton (COP = 6.09)	0.50 kW/ton (COP = 7.03)
Year service starts	2011	2016	2011
Year service ends	2015	2020	2020
Electricity used, kWh/yr	650,000	576,557	500,000
Demand, kW	325	288.3	250
Electricity cost, \$/yr	\$78,000	\$69,187	\$60,000
Demand charge, \$/yr	\$26,000	\$23,062	\$20,000
Capital cost	n/a	\$230,000 (RSMeans)	\$287,500 (25% premium)

Table 7 — Discounted Cash Flows Over Time

2011	2012	2013	2014	2015	2016	2017	2018	2019	2020
–$258,500	$27,074	$25,276	$23,598	$22,032	$165,296	$8,194	$7,663	$7,166	$6,702

Now let's do the same analysis using a SPP approach. We divide the capital cost of the efficient chiller by the reduction in annual energy cost ($104,000 – $80,000 = $24,000) to get

$$\text{Simple payback} = \frac{\$287,500}{\$24,000/\text{yr}} = 12.0 \text{ years}$$

Since the SPP is greater than the time frame of the analysis, this measure would not be accepted. However, the more comprehensive LCCA shows that the project has a positive NPV and makes financial sense.

Audit Reports

The primary deliverable of an energy audit is typically an energy audit report. The main purpose of audit reports is to recommend EEMs and provide clients with reliable information upon which they can base courses of action to save energy in their buildings. The audit report also provides background information on the site, its energy use, the systems in the building, and the process of the audit itself.

See Part 1 for specific requirements of Level 1, 2, and 3 audits. Within all of these levels of effort, the common elements of a good report include the following:

- Reliable recommendations for EEMs with
  - feasible measures and
  - reasonable and accurate energy savings and cost estimates.
- The owner/client has a clear path to follow.
- Existing systems are documented (consider including any completed PCBEA Sample Forms from Part 3 or www.ashrae.org/PCBEA).
- Proposed measures are communicated clearly, with sufficient detail for implementation.
- Supporting information is provided.
- Energy savings and cost summary are included.

Reports should be written in a manner that is easy to understand by the intended audience. The report may help to document efforts of site staff to reduce consumption or lower utility costs as well as any effects, positive or negative, of these efforts on the indoor environment. It may provide information on any major activities, savings to date, and future planned activities.

The successful energy audit meets the client's needs. Materials should be presented so they are usable and suited to the owner's level of expertise. An assessment that is left on the bookshelf is a failed endeavor. The energy audit does not end with the report—rather, the path to energy savings begins there.

Sample Outline

Typically, an energy audit report consists of the content outlined in the following list. While the detailed content and organization may vary, the elements discussed in the sections that follow are central to a comprehensive audit report.

Executive Summary

- Brief background and summary of scope
- EEM summary table
- Summary of benchmarking results

Background Information

- Contact information for stakeholders
- Audit scope, methodology
- Description of the site and the building
- Historical energy consumption and costs
- Benchmarking results
- End-use analysis results
- Client motivations or mandates satisfied by this effort

Description of the Existing Building Systems

- Occupancy
- Building envelope
- Lighting systems
- Mechanical systems
- Security, oversight, or operational requirements impacting this effort
- Photographs

Energy Efficiency Measures

- No-cost/low-cost measures
- Capital investment measures
- Demand response measures
- Distributed generation/renewable energy measures
- O&M measures
- Impact of current financial market, utility incentives, federal tax or grant support, etc.

Supporting Information

- Analysis
- Measured data or monitoring results
- Manufacturers' information or "cut sheets"

- Plans and sketches
- Additional information or specifications

For more detail on the content of each of these sections, see the following summaries.

Executive Summary

The executive summary should present the basic findings of the audit in a clear and concise manner. It should be suitable for nontechnical decision makers and include a brief description of the energy audit scope and a description of the site. The primary emphasis of the executive summary is a list of recommendations for the facility, including proposed project costs, energy and cost savings estimates, and the results of the economic analysis. These results are typically summarized in a table of proposed EEMs.

Also important is what the executive summary omits. The summary should be kept brief and not be bogged down with excessive discussion of technical approaches or project details. It needs to be free of extraneous information not pertinent to the overall findings. The emphasis should be on the findings and conclusions, not the methods of the investigation. It should present information about proposed projects that raise interest in the reader and promote the recommendations provided in the report.

Presenting Results (EEM Summary Table)

For decision makers, the primary value of an energy audit lies within the set of recommendations that, if implemented, will save energy and operating costs for the facility. The EEM summary table is the focal point for reviewing these recommendations in a brief form. In it the auditor summarizes the recommendations and their associated energy and cost savings. Finally, the audit report should suggest how, if implemented, the measures will satisfy organizational objectives or meet mandates.

The summary table typically includes the following minimum elements for each recommendation:

- EEM description
- Electricity and water consumption savings (kilowatt-hours, kilogallons, kilolitres)
- Peak demand savings (kilowatts)
- Fuel savings (British thermal units, therms, pounds of steam, megajoules, etc.)
- Operating cost savings (dollars)
- Implementation cost estimate (dollars)
- Financial evaluation (e.g., SPP, IRR, LCC, NPV)

The summary table may also include the following:

- O&M cost savings (dollars)
- Life-cycle measure cost (present or future worth)
- Discounted payback period

- Utility incentives
- IRR
- ROI
- Recommendations for demand response measures
- Equipment useful life (years)
- Recommendations for self generation (cogeneration, photovoltaics, etc.)
- Carbon emissions
- Cumulative savings totals

Supporting information with the summary table might include the following:

- Utility incentive rates
- Utility rate summary

An example EEM summary table is shown in Figure 34 and included in Part 3. This example is a Microsoft Excel® spreadsheet that includes formulas for calculating financial evaluation criteria based on user-supplied costs and savings. The focus of most decision makers is the bottom line, usually the simple payback of a project, as shown in Figure 35.

Background Information

This section of the report discusses the state of the building as it exists at the time of the energy audit. It summarizes information that is key to understanding the site, its existing energy use, the capital improvement plan, costs, systems, and controls.

The depth of the background information section can be tailored to the client's needs. For many building owners and operators, documentation of existing systems is sparse, so a thorough equipment inventory, with a description of current control strategies, is very useful. For other clients, this information may

Measure Number	Measure Description	Annual Energy and Cost Savings				Payback		Payback with Incentive				
		Peak Savings (kW)	Electricity Savings (kWh)	Gas/Fuel Savings (therms)	Total Cost Savings	Measure Cost	Simple Payback (yr)	Potential Utility Incentive	Net Measure Cost	IRR (over Life of Measure)	NPV	Simple Payback (yr)
EEM-1	Replace incandescent lamps with CFLs	7.6	15,245	-	$ 1,906	$ 1,875	1.0	$ 545	$ 1,330	132%	$ 3,958	0.7
EEM-2	Reduce Pressure Setting on Pneumatic Compressor	-	2,312	-	$ 206	$ -	-	$ -	$ -	N/A	$ 571	-
EEM-3	Install VFD on tenant condenser loop pump to reduce flow	19.0	163,872	-	$ 25,188	$ 17,386	0.7	$ 13,110	$ 4,276	589%	$ 200,021	0.2
EEM-4	Install Variable Frequency Drive and Implement Demand Control Ventilation for AHU-1	-	12,448	423	$ 2,290	$ 11,136	4.9	$ 1,418	$ 9,718	20%	$ 8,855	4.2
EEM-5	Install CO Sensors and VFD to Control Garage Exhaust Fan	18.8	48,948	-	$ 8,811	$ 25,616	2.9	$ 3,916	$ 21,700	39%	$ 49,762	2.5
EEM-6	Add Hot Water Resets Control and Install Condensing Boiler	-	44,838	17,203	$ 25,274	$ 102,511	4.1	$ 20,790	$ 81,721	31%	$ 261,758	3.2
EEM-7	Replace Garage HPS Fixtures With LED Fixtures With Integrated Motion Sensor	6.8	29,854	-	$ 4,114	$ 29,598	7.2	$ 2,174	$ 27,423	12%	$ 18,316	6.7
EEM-8	Recommission Economizers on All Air Handlers	5.0	22,342	-	$ 2,904	$ 12,864	4.4	$ 2,511	$ 10,354	12%	$ 2,577	3.6
TOTALS (Recommended Measures)		57.3	339,860	17,625	$ 70,692	$ 200,985	2.8	$ 44,464	$ 156,521	44%	$ 545,819	2.2

Figure 34 — Sample EEM Summary Table

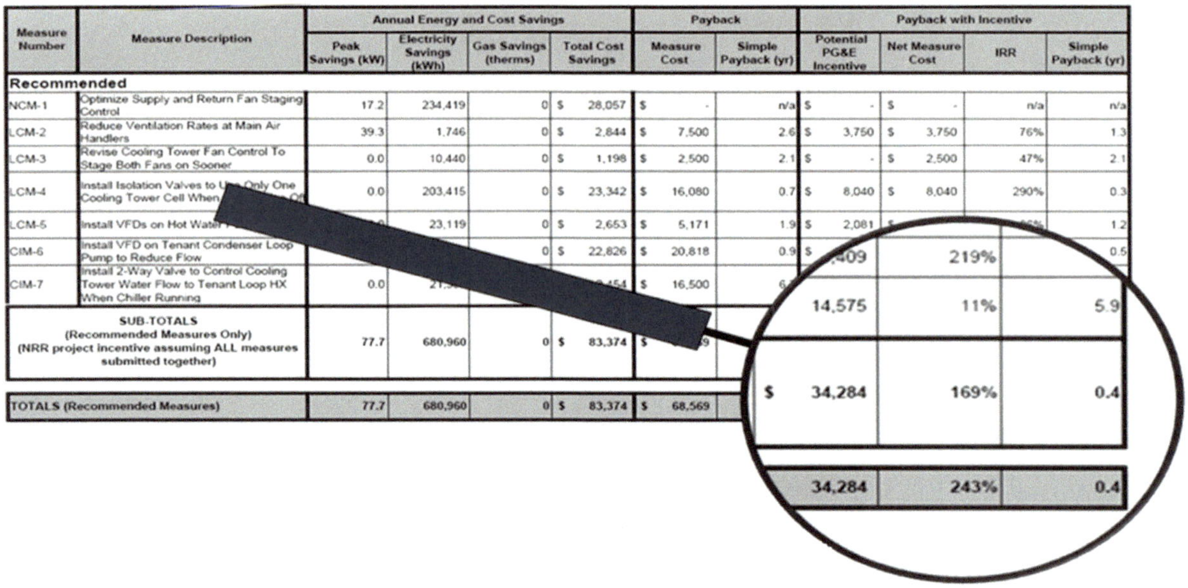

Measure Number	Measure Description	Annual Energy and Cost Savings				Payback		Payback with Incentive			
		Peak Savings (kW)	Electricity Savings (kWh)	Gas Savings (therms)	Total Cost Savings	Measure Cost	Simple Payback (yr)	Potential PG&E Incentive	Net Measure Cost	IRR	Simple Payback (yr)
Recommended											
NCM-1	Optimize Supply and Return Fan Staging Control	17.2	234,419	0	$ 28,057	$ -	n/a	$ -	$ -	n/a	n/a
LCM-2	Reduce Ventilation Rates at Main Air Handlers	39.3	1,746	0	$ 2,844	$ 7,500	2.6	$ 3,750	$ 3,750	76%	1.3
LCM-3	Revise Cooling Tower Fan Control To Stage Both Fans on Sooner	0.0	10,440	0	$ 1,198	$ 2,500	2.1	$ -	$ 2,500	47%	2.1
LCM-4	Install Isolation Valves to [illegible] Only One Cooling Tower Cell When [illegible]	0.0	203,415	0	$ 23,342	$ 16,080	0.7	$ 8,040	$ 8,040	290%	0.3
LCM-5	Install VFDs on Hot Wat[illegible]	[illegible]	23,119	0	$ 2,653	$ 5,171	1.9	$ 2,081	[illegible]	[illegible]	1.2
CIM-6	Install VFD on Tenant Condenser Loop Pump to Reduce Flow	[illegible]	[illegible]	0	$ 22,826	$ 20,818	0.9	$ [illegible]	[illegible]409	219%	0.5
CIM-7	Install 2-Way Valve to Control Cooling Tower Water Flow to Tenant Loop HX When Chiller Running	0.0	[illegible]	[illegible]	[illegible]	$ 16,500	[illegible]	[illegible]	14,575	11%	5.9
SUB-TOTALS (Recommended Measures Only) (NRR project incentive assuming ALL measures submitted together)		77.7	680,960	0	$ 83,374	$ [illegible]	[illegible]	[illegible]	$ 34,284	169%	0.4
TOTALS (Recommended Measures)		77.7	680,960	0	$ 83,374	$ 68,569	[illegible]	[illegible]	34,284	243%	0.4

Figure 35 — Many Decision Makers Focus on the Bottom Line—Simple Payback and IRR

already be summarized in a systems manual at the site, making further documentation redundant. Energy audit costs can be kept lower by requiring less effort in this regard. Green building rating systems may require this documentation in some cases, so if omitting the inventory for cost savings is considered, the scope should be discussed prior to the project's start. Table 8 includes key background information and the descriptions that are often included in this section of the report.

Description of the Site and the Building

This section of the report includes information on the building envelope, lighting, mechanical, and any other significant energy-using systems. At a minimum it should include the following:

- Floor area of the conditioned space (may be summarized by space type, e.g., office, retail, industrial, school, hospital, data center, etc.)
- Occupancy—number of occupants and hours of operation
- Envelope—descriptions of walls, fenestration, and roof
- Lighting—existing lighting fixtures and controls
- Mechanical—existing heating, cooling, and ventilation equipment and controls
- Other—compressed air systems, office equipment, computer equipment, or any element that makes up a significant fraction of the facility's energy use

The level of detail for each of these sections may be tailored to the client's needs and the potential for energy savings. For example, for the building shown

Table 8 — Background Information to Include in an Audit

Background Information	Description
Contact Information	Provides essential contact information for stakeholders for follow up and project implementation. Include at a minimum the owner's representative, the site contact, and the energy auditor. May also include utility account representative, vendors, contractors, and other associated parties.
Audit Scope	A brief summary of the level of effort at the site. May also describe analytical methods such as calculation or modeling approaches.
Description of the Site and the Building	Summary of the existing systems and controls in the building. See the subsection Description of the Site and the Building for more detail.
Historical Use	Tables and/or graphs summarizing the historical energy and water use at the site. See the subsection Historical Energy Consumption and Costs for more detail.
Benchmarking	Describes how the building compares to similar building stock. Methods can vary to correct for building size, climate, type of space (office, school, etc.), occupancy, and other variables. Buildings may be compared with similar types or within a portfolio belonging to the client (as in a chain retailer).
End-Use Breakdown	The energy balance displays the auditor's estimate of the end-use breakdown for the facility as a percentage of the historical energy use.

in Figure 36, the description of the building should include details about the glazing that makes up such a high percentage of this building envelope.

The conclusions and related results of previous studies may also be included in this section.

As an alternative to a building systems description in the audit report, the energy assessment team may provide a systems manual as a separate deliverable. A systems manual is a full description of the primary systems of the building, including design intent and intended control strategies. A systems manual is typically delivered as part of a retrocommissioning study; however, it is a valuable component of any energy study because it provides a reference for facility staff.

The energy auditor should alert the building owner and appropriate site staff to any unsafe conditions observed at the site. In addition, the auditor should call attention to any system or condition that does not meet applicable codes for the jurisdiction.

Historical Energy Consumption and Costs

Reviewing the historical energy use at the site (Figure 37) is an essential element of a successful energy audit. One cannot make energy-saving recommendations without first understanding the energy use of the existing facility.

The section on historical energy use typically includes tables and/or graphs that summarize the following:

- Electricity consumption (kilowatt-hours or kilowatt-hours per day)
- Peak demand (monthly peak kilowatt)

Figure 36 — The Audit Report Description of this Building Should Include the High Window-to-Wall Ratio on this Façade

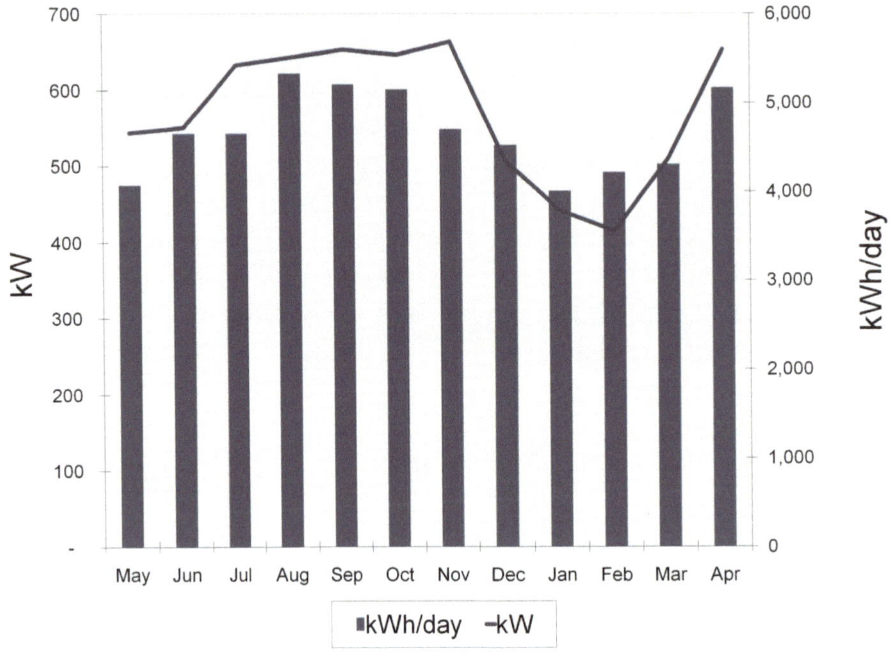

Figure 37 — Sample Electricity Consumption Data

- Fuel consumption (British thermal units, therms, megajoules)
- Energy costs (dollars or dollars/day)
- Daily load profiles

The historical annual energy consumption by fuel type is typically presented for at least one year (or for more years if it is available).

This information is quicker to review if it is presented graphically for the owner. It is common to normalize consumption data for the number of days in the month if the data comes directly from energy bills. This corrects for any apparent discrepancy that might arise simply because there were more or less days in the billing cycle.

Historical data may also include interval data, showing consumption or peak kilowatt-hour for 15-minute periods of the day. This data is now increasingly available for commercial accounts from electric and gas utilities who are installing smart meters (see Figure 38 for an example of hourly demand data captured by a smart meter). This data may be available up to the minute or to the next day. This interval data can be very valuable in identifying persistent nighttime loads or specific building responses that occur regularly or are weather dependent.

This section should also include a brief description of applicable utility rates. Care should be taken by the audit team to identify rate thresholds that might be reached (e.g., a reduction in peak demand forcing the facility to apply to a lower demand rate and often a higher cost per kilowatt-hour).

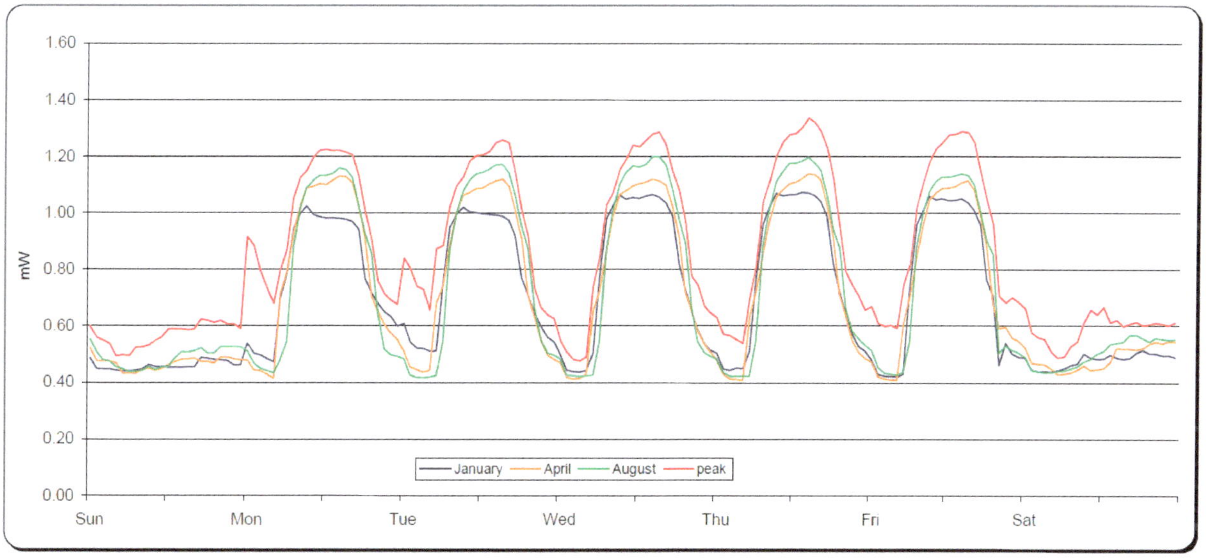

Figure 38 — Sample Interval Billing Data—Hourly Demand by Day of the Week and Month

Description of the Existing Building Systems

Photographs

Digital cameras are an essential component of any energy auditor's tool kit. Figure 39 shows an example of a photograph that can be used to alert the building owner about significant maintenance issues.

Digital cameras are also very valuable to the auditor for recording information on site. In many cases it is easier, faster, and safer to photograph existing equipment nameplates than to lean over equipment or record data while atop a ladder (see Figure 40).

Energy Efficiency Measures

The main body of the audit report should focus on the recommendations for the facility. This section should include the results of interviews, analyses, and measurements from the site. The auditor should develop a list of individual EEMs with individual project descriptions.

For each recommended EEM, the audit report may include the following:

- Observation of existing conditions
- Recommendation for changes

Figure 39 — An Audit Photograph Documenting Maintenance Issues such as this Inoperable Fresh Air Damper Might Spur the Client to Take Corrective Action

- Basis for how the measure will save energy
- Basis of assumptions for costs
- Key analytical assumptions
- Performance specifications for recommended equipment
- Key control changes necessary to achieve savings
- Any operational changes required of staff to support proposed measure

EEMs may range from simple, nontechnical measures (no-cost solutions) to those needing major investment and possibly more detailed (Level 2) assessment with an economic analysis and/or design.

No-cost/low-cost measures require minimal investment and can often be implemented without further study. These may include general good-housekeeping practices, "turn-off" campaigns, avoiding wasteful practices, adjustment of existing controls to match actual requirements of occupancy, or installation of small items (e.g., thermostats, insulation of piping sections, fixing cracks in window frames, etc.).

Capital investment measures are expected to need further detailed Level 2 analysis. These measures may require an executive level of approval and budgeting prior to implementation. Such measures may include building wall insulation; replacement or upgrading of a plant and equipment, such as boilers or chillers; or installation of controls.

Supporting Information

Additional reference information or supporting calculations may be included in the report to aid in the implementation of measures and to inform the building owner. This supplemental information may be best suited for an

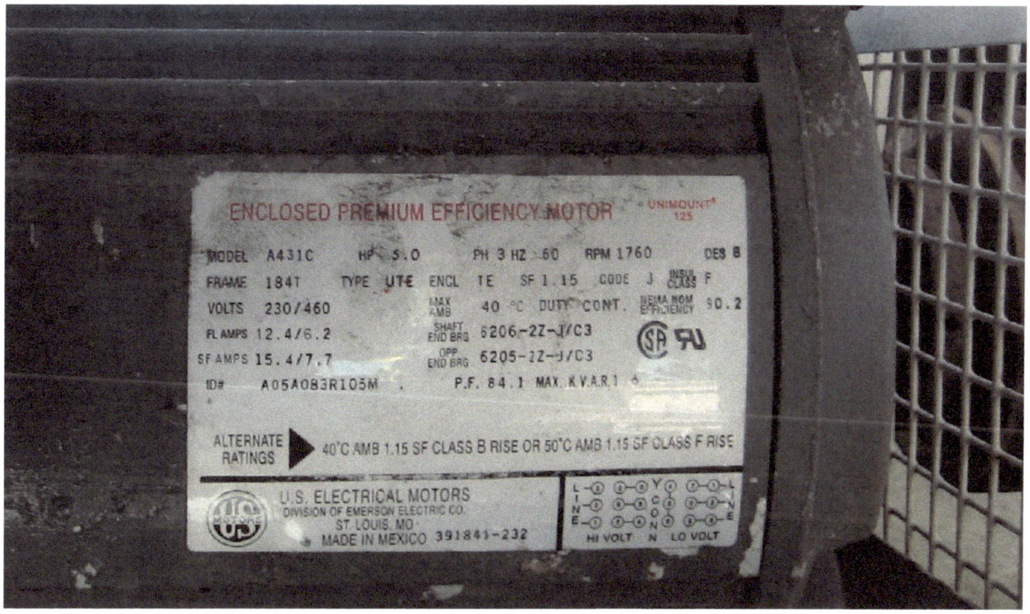

Figure 40 — Reading this Nameplate from a Digital Photograph is Easier and Safer than Leaning over the Motor

appendix at the end of the report so that the audit remains focused. Examples of additional materials include the following:

- Supporting analysis
- Equipment inventories
- Manufacturers' performance data or "cut sheets" for suitable equipment
- Performance specifications
- Monitored data
- Field data sheets

Quality Control

Quality control (QC) typically involves reviewing the energy audit report and calculations to ensure that recommendations are reasonable and feasible and that the analysis is based on solid engineering foundations and is without errors that might affect the conclusions. QC may be performed by someone on the energy auditor's team, by a third party or peer reviewer, or by the client. In any of these cases, QC is an important part of the process to ensure that recommendations are well founded and that the conclusions are apt.

The QC reviewer should pay special attention to the following potential issues when reviewing the report:

- The proposed measures are feasible and appropriate for the building.
- The proposed measures meet applicable building codes.
- The data are internally consistent.
- The savings estimate methods follow established principles and methods.
- The estimates of potential energy savings are reasonable compared to quick estimates and historical energy use.
- The proposed cost estimates are reasonable relative to field experience.

What to Ask for in an Audit Report

- Actionable recommendations
- Realistic treatment of rates
- Transparent analysis
- Guidance to more resources
- Reasonable savings estimates
- Reasonable cost estimates
- Analysis of interactive effects
- Measurements of key input variables
- Monitoring of baseline performance
- Hourly modeling

- The interactions between EEMs are identified and addressed (though sometimes not quantified).
- The recommendations and report meet the project scope, goals, and client's needs.
- Any financial discussion includes current and viable mechanisms available per the tax structure, location, and motivations of the client.

Follow-Up Meeting

A follow-up meeting can be a good way to ensure the effectiveness of an energy assessment by discussing recommendations and formulating an implementation plan. In this meeting, the audit team and the client can review the key points of the audit report. The report can be supported effectively if the assessor listens and responds to the client's questions and concerns.

The purpose of the follow-up meeting is to inform the client of the team's findings, make sure that the client's goals were met, address concerns that are raised, and develop a plan for moving forward. It is all too common for energy audit clients to be pleased with the auditors work but then fail to implement the recommendations. The follow-up meeting can help build teamwork among the relevant stakeholders (the owner, financial decision makers, site maintenance staff, and the auditing group) and keep the project momentum going by beginning an implementation plan while the partners are engaged and informed.

If the client has a group of decision makers, a presentation may be appropriate to illustrate some of the findings. At a minimum, the presentation should review the audit team's methods and the primary findings and leave plenty of time for questions and feedback. Slides are most effective if they are evocative, include figures, photographs, and data, and are not a simple list of bullet points. The presenter should also be prepared to lead a meeting without slides, as projector lamps can fail or meeting rooms may be too bright. Stating "let's skip the slides" may even impress a client who is tired of presentations.

What Next—Implementing Measures

Overview

Implementation begins once the owner has selected specific EEMs to pursue and has secured the necessary funding to see them through completion.

Implementation of the selected measures depends on the client's resources, plans, and priorities. The assessor may be asked to join the team to provide assistance with the design, implementation, verification, and Cx of measures, or the owner may proceed without further consultation from the assessor. The involvement of the assessor may be influenced by the procurement path for the energy audit and implementation, as described below.

Several contracting options are commonly used for implementing efficiency measures, each with their own pros and cons. The design and installation could be performed by the client's in-house staff or contracted to a design-build or traditional plan and specification firm. The benefits and downfalls of each option depend on the chosen measures and the capability of the installing party. In-house staff can provide a cost-effective way to implement measures if they have the appropriate skill set. This may avoid lengthy contract negotiations, contractor mark-ups on materials, and expensive labor rates for electricians or controls technicians. However, if specialized personnel are not available on staff, using in-house staff for implementation may result in unsuccessful installation.

A design-build approach may reduce the initial cost of producing construction documents for a contractor. This can add up to be a significant cost savings to the client if the firm has adequate experience with the implementation that is required. Without the correct experience, a design-build firm will not likely be successful in implementing the measures if they have partial or no design documents or insufficient guidance.

A traditional plan and specification firm coupled with an installing contractor helps ensure that measures are designed correctly and installed properly. Under this option, construction documents, including specifications and plans, are produced by specialized personnel. The contractor implements each measure based on these documents. However, depending on the measure, producing these documents may require a significant amount of time and cost.

The use of an energy service company, commonly known as an ESCO, is another option for implementing EEMs. The National Association of Energy Service Companies (www.naesco.org) defines an ESCO as a business that develops, installs, and arranges financing for projects designed to improve the energy efficiency and maintenance costs for facilities over a seven- to twenty-year time period. An ESCO is different from consulting firms and contractors because the company's compensation, and often the project's financing, is directly linked to energy savings. The company's role will depend on the measures to be implemented, the assessor's capabilities and experience, and the needs and capabilities of the owner. The availability of capital through an ESCO performance contract can be attractive, particularly with organizations such as governments and universities where investment capital is hard to secure. However, writing a successful ESCO contract can be a complicated process that requires experience in energy efficiency, construction, and financing. Guidance from someone experienced with performance contracting with ESCOs is important in designing an agreement that achieves the desired end and is beneficial to all parties.

Implementing Energy Efficiency Measures

The successful implementation of EEMs is a result of coordination between the client, the auditor, designers, and contractors throughout the duration of the project. Figure 41 illustrates measurable energy savings as a result of successful implementation of EEMs. Communicating appropriately and in a timely manner will help to resolve problems that might arise during the implementation phase.

In some ways, implementation of efficiency measures is just like any other construction project and brings many of the same issues to the table. The proj-

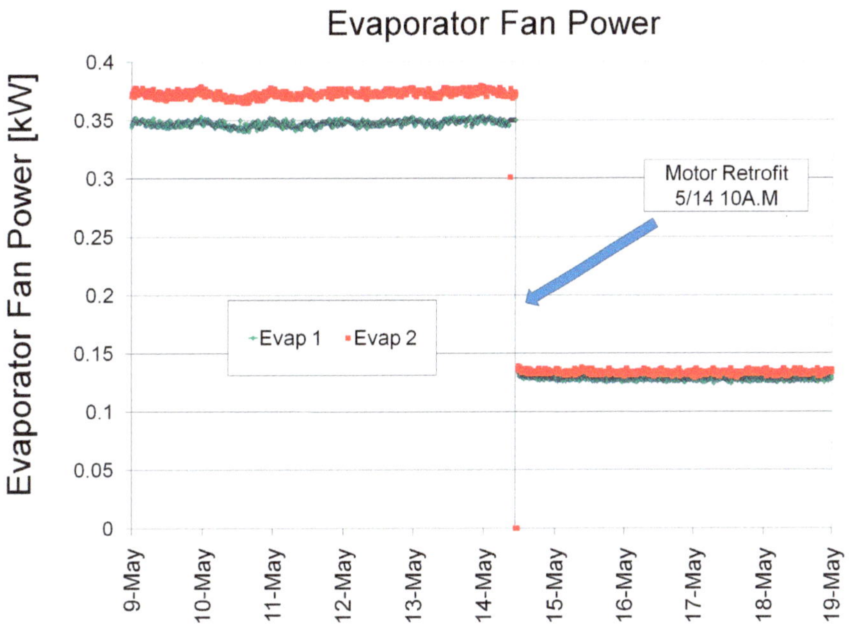

Figure 41 — Successful Implementation Results in Measurable Energy Savings

ects are typically retrofit in nature and are often relatively small, especially when taken individually. Small retrofit projects add their own level of complexity to the procurement and construction process and can tax the abilities of even the most sophisticated owners.

There is, however, one very important distinction to be made between energy efficiency projects and other construction projects: *the primary purpose and goals of the project are to save energy and operating costs.* It is important to recognize this and make sure it is communicated and fully integrated into the entire implementation process. This is not to say that other criteria such as serviceability, comfort, and life aren't important; but the basis for doing the project is to save energy.

Beginning Implementation

Implementation tasks may vary significantly from measure to measure and depend on the scope of work included in the assessment, the owner's procurement processes, the estimated cost, design and construction administration requirements, the construction trades and contractors that will be involved, and, finally, who is impacted by the measure (i.e., the users).

The owner needs to determine what procurement path to follow and what level of due diligence is required. For larger projects, a traditional design-bid-build approach may be appropriate. Smaller projects could follow a streamlined approach that relies on a detailed performance-based scope of work rather than traditional drawings and specifications, with solicitations limited to preselected/qualified contractors. In some cases a design-build approach may make sense, especially when the measures are extremely complex and difficult to define in a way that could lead to truly competitive pricing.

Often, the procurement path is dictated by the owner's organizational procurement policies and there is little flexibility. Regardless of the path chosen or dictated, it is important to recognize the differences between more traditional construction projects and energy efficiency projects—the latter are driven by energy savings (i.e., performance) and often, but not always, require specialized services for both design and construction.

For all but the smallest projects, the owner should consider forming a defined team and assigning tasks and responsibilities to each member. Potential members include an owner's project manager, the assessor, designers, financial and/or procurement specialists, the users, contractors, and Cx agents. In many cases, multiple team roles can be filled by one person or firm. All team members must be made aware of the primary purpose of the project—*to save energy*.

Scope of Work

Adequately defining the scope of work is a critical task for the energy assessor during the implementation phase. Sometimes building an adequate scope of work will require additional investigation or the assistance of a specialized discipline, such as a mechanical design engineer or controls designer. Often there can be a big difference between a scope of work and a procurement document. Depending on the owner's requirements, creating an appropriate procurement document may require preparation of design drawings and/or detailed specifications. In some cases a detailed performance-based scope might be sufficient.

Selecting a Contractor

It is important to select a contractor who understands the implementation requirements and has the knowledge and experience to successfully implement the measure. In addition, it is important that the installing contractor is informed of the energy goals of the project and any verification methods that may require ongoing contractor involvement. If an energy performance procurement path was chosen at the beginning of the project, the contractor may be included in the assessment and implementation team at the initial awarding of the contract.

Construction Management

Construction project management may benefit from planning and coordination by the energy auditor to ensure that the client's requirements are met and the EEM is implemented successfully. During the implementation process, regular coordination is generally required between the client, contractors, design personnel, and the energy auditor to keep the project on budget and on schedule. Construction management responsibilities of the energy auditor include reviewing plans, specifications, and costs to ensure that the intent of the EEM is not lost and that the performance specifications are maintained.

Measurement and Verification

Before implementation work begins, procedures should be established to validate the EEM installation and operation. Verifying the estimated energy savings may be important to the client or may be required by other parties, such as the utility provider.

Typically the guidelines and methods used for verifying energy savings will follow or be based on the IPMVP (EERE 2002) as well as ASHRAE Guideline 14 (ASHRAE 2002). The IPMVP is a guidance document that provides framework for measuring, estimating, and reporting savings achieved by implementing EEMs. ASHRAE Guideline 14 is a reference for calculating the energy and demand savings using measurements.

Selecting an appropriate M&V method depends on several factors. These include, but are not limited to, the following:

- Implementation costs and estimated savings
- Complexity of the EEM or system
- Interaction of EEMs within a facility
- Level of certainty for savings

Verification may include frequent spot measurements, data obtained from the utility, or additional equipment-level metering. Metering can use sophisticated devices that automatically read and transmit data to a central location or less expensive metering devices that require regular readings by building maintenance and/or security personnel.

Costs for automatic metering devices, such as adding points to a DDC system, must be weighed against the benefits. Many energy managers find it helpful to collect energy consumption information hourly. The energy manager should review data while they are current and take immediate action if profiles indicate a trend in the wrong direction. These trends could be caused by uncalibrated controls, changes in operating practices, or mechanical system failures, which should be isolated and corrected as soon as possible.

During the monitoring process, problems that are identified can be immediately addressed and fixed. This will reduce unnecessary energy consumption and allow the building to achieve persistent energy savings.

Part 3

Resources for Conducting Energy Audits

Sample Forms and Templates

This section includes sample energy audit forms and templates to assist building energy auditors in collecting the data required to complete comprehensive energy and financial analyses of proposed modifications to a building. The samples provided here include data collection forms for components of a commercial building that contribute to heating and cooling loads or consume energy as well as templates for completing the Preliminary Energy-Use Analysis (PEA), end-use breakdowns, and energy efficiency measure (EEM) summary tables.

The samples included here, as well as additional forms and resources not reproduced in this book, may be downloaded from www.ashrae.org/PCBEA and used as the bases for data collection and reporting results from energy audits. The forms and templates available on the Web site are in Microsoft Excel® spreadsheet format. As these forms and templates may be revised from time to time, users should refer to the Web site for the most up-to-date version when using the resources for their audit projects. When a template is revised, it will be posted to www.ashrae.org/PCBEA with a new date at the end of the file name; this revised form will replace any versions of the template previously available on the Web site.

Sample Forms

PCBEA Sample Forms

The following information and data collection forms, reproduced in this section, are available in the spreadsheet "ASHRAE PCBEA Sample Forms.xls" at www.ashrae.org/PCBEA.

- Intro
- TOC
- 1.0 Basic Site Information
- 1.10 Capital Improvement PLan
- 1.11 Operations and Maintenance Costs
- 1.12 Space Function Summary
- 1.31 Schedules—Occupancy
- 1.51 Interior Lighting
- 1.52 Exterior Lighting
- 1.81 Domestic Hot Water—Equipment
- 1.82 Domestic Hot Water—Fixtures and Use
- 2.0 Boilers
- 2.1 Chillers
- 2.2 Cooling Towers and Fluid Coolers
- 2.3 Pumps and Piping Systems
- 2.4 Air-Handling System Equipment
- 2.5 Air-Handling System Controls
- 2.10 Packaged Units: DX, Heat Pumps

The Table of Contents of the PCBEA Sample Forms lists the many other sample data collection forms available in the spreadsheet.

Procedures for Commercial Building Energy Audits, Second Edition

Sample Form Instructions

Purpose

Gathering the right data, with the appropriate amount of detail, is a key component to realizing the maximum savings from an energy audit. However, important data can sometimes be overlooked or the level of detail may be insufficient to support energy and financial analyses. To address these issues, we have created these template data collection forms. The purpose of these sample forms is to assist building energy auditors in collecting data required to complete comprehensive energy and financial analyses of proposed modifications to buildings. These forms are best suited to Level 2 and Level 3 energy audits.

It is not necessary to complete these forms in entirety to perform a comprehensive energy audit of a commercial building—it is up to the auditor's judgment to collect information in sufficient detail to develop recommendations for a facility. The intent of the forms is to guide the auditor to collect data that they may need later, not to dictate the level of detail needed.

Acknowledgments

Rocky Mountain Institute, Integral Group, Taylor Engineering, kW Engineering

Sample Form Recommended Use

- Identify the information in these sample forms that is needed for your project.
- Identify your anticipated source for each piece of data. Typical data sources are listed below.
- Use the "citation" column where available in case you need to clarify the information at a later time.
- Use the "Sample Form Guidance" information provided for some forms as examples of the type of information you should be collecting; you may want to print them out and take them with you to the site.
- Collect as much data as possible prior to a site visit.
- Print the Sample Forms (sometimes multiples of each) and take them with you to the site (alternatively, take electronic copies).
- You may need to change the column widths or margins to print using your computer and printer.

Typical Data Sources

- Facility manager
- Q&A with building occupants
- As-built drawing sets
- Utility bills
- Interval energy data
- Utility rate structures, purchase agreements
- BAS trend data
- On-site spot measurements
- Temporary logger data

Expert Tip

When using these forms in the field, put a star next to any energy efficiency measure opportunity you see. This could include equipment replacement options, control improvements, or O&M repairs that might lead to an audit recommendation. This will help highlight items for later analysis.

Tools for a Level 1 Audit

- Identification
- Notebook/clipboard
- Phone
- Audit forms appropriate to the site
- Digital camera
- Multi-tool
- Flashlight
- Digital thermometer and/or humidity meter
- Hearing protection (for mechanical rooms)

Tools for Level 2 and Level 3 Audits

- Energy audit sample forms (from this book)
- Illuminance meter (light meter)
- Infrared temperature spot meter
- Electronic ballast spinner/detector
- Power meter and/or data loggers
- Temperature loggers
- Light loggers
- Motor runtime status loggers
- Tape measure

Specialized equipment for detailed assessments may also include

- Boiler stack gas analyzer (for combustion tests)
- Ultrasonic flowmeters (for monitoring chilled-water, condenser-water, or heating-water flows)
- Infrared camera
- Pipe caliper
- Centralized data logging systems
- Personal protection equipment (PPE) per OSHA (may include construction hard hat, safety glasses, gloves)

TABLE OF CONTENTS

PCBEA Sample Forms

1.0 BASIC SITE INFORMATION

PCBEA Sample Forms

Project Name

Client Name

Site Address

City/State

Year

Building Type and/or Functions

Gross Area

Lease Type

No. Stories

Key Contacts

Name and Position	Phone/Email

Economic Criteria for Energy Projects

Previous Audit or Engineering Study Availability

Notes

NOTE: Request utility/fuel bills for two- to three-year period.

1.10 CAPITAL IMPROVEMENT PLAN

PCBEA Sample Forms

To calculate the marginal capital costs of efficiency measures, it is necessary to document any plans for capital improvement. Since a whole-building retrofit may alter the course of almost any type of planned improvements, it is important to document them all (several types are indicated below). Print multiple sheets as necessary.

Type

HVAC

Description of planned replacement, renewal, or system expansion		Planned	Cost	Citation

Envelope

Description of planned replacement, renewal, or system expansion		Planned	Cost	Citation

Electrical

Description of planned replacement, renewal, or system expansion		Planned	Cost	Pump hp

Plumbing

Description of planned replacement, renewal, or system expansion		Planned	Cost	Citation

Aesthetic

Description of planned replacement, renewal, or system expansion		Planned	Cost	Citation

Aesthetic

Description of planned replacement, renewal, or system expansion		Planned	Cost	Citation

Other

Description of planned replacement, renewal, or system expansion		Planned	Cost	Citation

TOTAL $ -

1.11 OPERATIONS AND MAINTENANCE COSTS

PCBEA Sample Forms

Documenting annual operations and maintenance costs will allow you to calculate changes in these costs as a result of the retrofit. If possible, gather at least three years of data. Print additional sheets for each year.

Year: ________

Description	Cost	Citation / Receipt No.
1		
2		
3		
4		
5		
6		
7		
8		
9		
10		
11		
12		
13		
14		
15		
16		
17		
18		
hp		
20		
TOTAL	0	

1.12 SPACE FUNCTION SUMMARY

PCBEA Sample Forms

Use this sheet to allocate the gross floor area of your building into standard space function types. This can be used to determine a weighted Energy Utilization Index (EUI) for the building based on standard space function types.

Dropdown options are provided for consistency with ENERGY STAR® space types. If you chose to benchmark with ENERGY STAR data, additional information may be required.

#	Space Function Type	Gross Floor Area	Weekly Operating Hours	Weeks/ Year	# Occupants	# PCs	Principal Lighting Type	Principal HVAC Type	% of Spaces Heated	% of Spaces Cooled
1										
2										
3										
4										
5										
6										
7										
8										
9										
10										
11										
12										
13										
14										
15										
16										
17										
18										
19										
20										
Total		0			0	0				

1.31 SCHEDULES—OCCUPANCY

PCBEA Sample Forms

Fraction of Peak Occupancy

Zone Name/Description								
Days	M	T	W	TH	F	Sat	Sun	Hol
12–1 am								
1–2 am								
2–3 am								
3–4 am								
4–5 am								
5–6 am								
6–7 am								
7–8 am								
8–9 am								
9–10 am								
10–11 am								
11–12 pm								
12–1 pm								
1–2 pm								
2–3 pm								
3–4 pm								
4–5 pm								
5–6 pm								
6–7 pm								
7–8 pm								
8–9 pm								
9–10 pm								
10–11 pm								
11–12 am								

Drawings/Notes

Use this space to make drawings or take notes.

1.51 INTERIOR LIGHTING

PCBEA Sample Forms

Use this sheet to describe the electric lighting plan for each zone in your building. Print multiple sheets as necessary.

Zone Name/Description		Area (ft^2):

Fixture				
Fixture Type	Area Task Exit Track Display Other:		No. of Fixtures:	
Mounting	Recessed Suspended Surface Mount Other:		No. of Lamps/Fixtures:	
Lamp Type	Fluorescent Tube Length:	Diameter: T12 T8 T5 Other:	Wattage of Lamps:	
	Circle one: CFL Incandescent Halogen LED HID Other:			
Ballast	Circle one: Magnetic Electronic	No. of Ballasts/Fixtures:	BF:	Watts:
Control Type (circle one)	Manual Switch Bi-Level Switch Time Clock Daylight Sensor Occupancy Sensor Dimming Step Dimming None Other:		Total Watts:	

Work Plane Height: *Horizontal Illuminance:* *Vertical Illuminance Sensitivity:* *Lamp Color Temp:*

Fixture				
Fixture Type	Area Task Exit Track Display Other:		No. of Fixtures:	
Mounting	Recessed Suspended Surface Mount Other:		No. of Lamps/Fixtures:	
Lamp Type	Fluorescent Tube Length:	Diameter: T12 T8 T5 Other:	Wattage of Lamps:	
	Circle one: CFL Incandescent Halogen LED HID Other:			
Ballast	Circle one: Magnetic Electronic	No. of Ballasts/Fixtures:	BF:	Watts:
Control Type (circle one)	Manual Switch Bi-Level Switch Time Clock Daylight Sensor Occupancy Sensor Dimming Step Dimming None Other:		Total Watts:	

Work Plane Height: *Horizontal Illuminance:* *Vertical Illuminance Sensitivity:* *Lamp Color Temp:*

Fixture				
Fixture Type	Area Task Exit Track Display Other:		No. of Fixtures:	
Mounting	Recessed Suspended Surface Mount Other:		No. of Lamps/Fixtures:	
Lamp Type	Fluorescent Tube Length:	Diameter: T12 T8 T5 Other:	Wattage of Lamps:	
	Circle one: CFL Incandescent Halogen LED HID Other:			
Ballast	Circle one: Magnetic Electronic	No. of Ballasts/Fixtures:	BF:	Watts:
Control Type (circle one)	Manual Switch Bi-Level Switch Time Clock Daylight Sensor Occupancy Sensor Dimming Step Dimming None Other:		Total Watts:	

Work Plane Height: *Horizontal Illuminance:* *Vertical Illuminance Sensitivity:* *Lamp Color Temp:*

Zone W/ft^2		**Corresponding Schedule**	

Notes

1.52 EXTERIOR LIGHTING

PCBEA Sample Forms

Use this sheet to describe the electric lighting plan for the exterior of the building.
Print multiple sheets as necessary.

Zone Name/Description		Area (ft^2):

Manufacturer/Model				
Use Type	Security Advertising Parking Lot Parking Garage Bldg Façade Landscape Other:		No. of Fixtures:	
Mounting	Pole Bollard Building Facade Ceiling Other		No. of Lamps/Fixtures:	
Lamp Type (circle one)	Fluorescent or Neon Tube	Length:	FL Diameter: T12 T10 T8 T5 Neon Diameter (mm):	Wattage of Lamps:
	CFL Halogen LED HPS MH MV Incan Other:			
Ballast	Circle one: Magnetic Electronic	No. of Ballasts/Fixtures:	BF:	Watts:
Control Type (circle one)	Manual Switch Photocell Time Clock Photocell/Time Clock Motion Sensor Other:		Total Watts:	
Horizontal Illuminance:	*Vertical Illuminance Sensitivity:*	*Lamp Color Temp:*		

Manufacturer/Model				
Use Type	Security Advertising Parking Lot Parking Garage Bldg Façade Landscape Other:		No. of Fixtures:	
Mounting	Pole Bollard Building Facade Ceiling Other		No. of Lamps/Fixtures:	
Lamp Type (circle one)	Fluorescent or Neon Tube	Length:	FL Diameter: T12 T10 T8 T5 Neon Diameter (mm):	Wattage of Lamps:
	CFL Halogen LED HPS MH MV Incan Other:			
Ballast	Circle one: Magnetic Electronic	No. of Ballasts/Fixtures:	BF:	Watts:
Control Type (circle one)	Manual Switch Photocell Time Clock Photocell/Time Clock Motion Sensor Other:		Total Watts:	
Horizontal Illuminance:	*Vertical Illuminance Sensitivity:*	*Lamp Color Temp:*		

Manufacturer/Model				
Use Type	Security Advertising Parking Lot Parking Garage Bldg Façade Landscape Other:		No. of Fixtures:	
Mounting	Pole Bollard Building Facade Ceiling Other		No. of Lamps/Fixtures:	
Lamp Type (circle one)	Fluorescent or Neon Tube	Length:	FL Diameter: T12 T10 T8 T5 Neon Diameter (mm):	Wattage of Lamps:
	CFL Halogen LED HPS MH MV Incan Other:			
Ballast	Circle one: Magnetic Electronic	No. of Ballasts/Fixtures:	BF:	Watts:
Control Type (circle one)	Manual Switch Photocell Time Clock Photocell/Time Clock Motion Sensor Other:		Total Watts:	
Horizontal Illuminance:	*Vertical Illuminance Sensitivity:*	*Lamp Color Temp:*		

Zone W/ft^2		**Corresponding Schedule** (Reference Sample Form 1.32)	

Notes

1.81 DOMESTIC HOT WATER—EQUIPMENT

PCBEA Sample Forms

Type (circle one)	Storage Water Heater Instantaneous (Tankless) Boiler Tankless Coil Dishwasher Booster Indirect—Steam to Hot Water Indirect—Hot Water to Hot Water Purchased Hot Water Purchased Steam Heat Pump Solar Solar Backup Other:		
Storage Size (gallons, lb, …)			
Location(s)		**No. of Units:**	
Fuel Type	Natural Gas Electric Fuel Oil LPG Heat Recovery Other:		
System Integrity			
Manufacturer Information	Name and Model		
	Serial No.		
	Tank Capacity (if applicable)	gallons	
	Rated Input Capacity (circle units)	kBtu/h kW	
	Efficiency Rating (circle units)	Energy Factor Thermal Efficiency AFUE COP	
	Tank Internal Insulation R-Value (if applicable)	Tank Dimensions:	
	Age (years)	Condition: (describe)	
Does the tank have an external insulation jacket? Yes No R-Value?			
Average Hot-Water Temperature (°F)			
Are the Hot-Water Pipes Insulated? Yes No How thick?			
Loop Design	Design Flow (gpm)	**Pump hp**	
Recirculation Pump	Is there a recirculation pump? Yes No	Min Flow (gpm)	**Motor Eff**
	Pump Head (ft)	Max Flow (gpm)	
	Control (circle one): Continuous Temperature Timer Demand Other:		
	Control Settings (e.g., setpoint temperature):		
	Average Recirculation Time (min)		

Drawings/Notes
Use this space to make drawings or take notes about the hot-water equipment.

1.82 DOMESTIC HOT WATER—FIXTURES AND USE

PCBEA Sample Forms

Use this sheet to document the hot-water fixtures and their use. This will allow you to calculate hot-water heating load reductions resulting from more efficient fixtures.

Type (circle one)	Lavatory Faucet Kitchen Faucet Dishwasher Laundry Washer Shower Other:			
Location(s)			**No. of Units:**	
Manufacturer Information	Name and Model:			
	Serial No.:			
	Flow/Cycle Rate: (circle units)	gpm: gallons per cycle:		
	Age: (years)			
Does the Fixture Leak? Yes No Estimate how much:				
Fixture Condition Good Poor Describe:				

Use Rates (enter information where applicable)	Applicable to:		
	Dishwasher Laundry Washer Shower	Cycles per Day:	When do most cycles occur? Morning Midday Evening
	Lavatory Faucet Other Fixture	Average No. of Users per Day:	When does most use occur? Morning Midday Evening
	Kitchen Faucet Other Fixture	Average Time of Use per Day: (mins)	When does most use occur? Morning Midday Evening

Drawings/Notes
Use this space to make drawings or take notes about the fixture location or hot-water use.

2.0 BOILERS

PCBEA Sample Forms

General Information

Site Name			
Boiler ID/Tag			
Location/Service			
Mfr/Model #			
Year Built		Serial Number	
Condition of Tubes/Sections			
Condition of Combustion Chamber			

Burner

Type: Air Atomized / Pressure Atomized / Rotary Cup / Natural Draft / Power Gas

Mfr/Model #			
Year Built		Serial Number	

Boiler Type

Scotch Marine	
Fire Tube—Fire Box	
Cast Iron Sectional	
Water Tube	
Electric	
Natural Draft	
Forced Draft	
Condensing	
Other ________	

Boiler Fuel

Natural Gas	
Fuel Oil (grade)	
Propane	
Electricity	
Other ________	
Dual Fuel	

Heating Fluid

Water (HWS/HWR) °F	
Glycol?	yes / no
Glycol %	
Steam psi (LP/HP)	
Other ____________	
Summer/Winter psi?	

Combustion Fan Data

Quantity	
Manufacturer	
Full-Load hp	
Frame Size	
ODP/TEFC	
Full-Load rpm	
Fan Efficiency	
Motor Efficiency	
Full-Load Amps	
Drive Type	

Design/Measured Data

Boiler	Design	Measured
Input/Output (Btu)		
gpm		
WPD (ft)		
EWT/LWT (°F)		
Fan	Design	Measured
Volts		
Phase		
FLA		
kW		
PF		

Accessories

Draft Control	
Flue Gas Recirculation	
Oxygen Trim Control	
Stack Economizer	
Blowdown Economizer	

Piping Configuration

Primary-Secondary	
Primary-Only	
Variable Flow	

Temperature Control

LWT Setpoint (°F)	
EWT Setpoint (°F)	
Reset?	yes / no

If yes, describe: ________

Feedwater

Condensate Returned (Y/N)	
Feedwater Pump (Y/N)	
Feedwater Pump (hp)	
gpm	
WPD (ft)	

Controls

Describe in notes time-of-day scheduling, boiler staging, cycling controls, settings, and observations.

____

____

____

2.0 BOILERS

PCBEA Sample Forms

Combustion Efficiency Test Results

Conditions/Firing Rate			
Stack Temperature			
Room Temperature			
Net Stack Temperature			
% O2			
% CO2			
Stack Draft			
CO (ppm)			
Smoke			
% CE			

Additional Notes/Observations

2.1 CHILLERS

PCBEA Sample Forms

General Information

Site Name			
Chiller ID/Tag			
Location/Service			
Manufacturer/Model #			
Year Built		Serial Number	

Chiller Data

Air or Water Cooled	
Refrigerant Type	
Piping Arrangement	
Chilled-Water Glycol	Prop / Eth / None
Glycol %	
Capacity (tons)	

Design/Operating Conditions

	Design	Measured
kW/ton @ 100% load		
kW/ton @ 50% load		
EER at ARI Conditions		
NPLV / IPLV		
Evap gpm		
Evap EWT/LWT (°F)		
Evap WPD (ft)		
Cond gpm		
Cond EWT/LWT (°F)		
Cond WPD (ft)		

Compressor Data

Quantity	
Type	
RLA	
Volts	
Phase	
hp	
VFD	
Compressor Stages	

Temperature Control

Cond Water Supply Setpoint	
Chilled-Water Supply Setpoint	
Computerized / Electronic / Pneumatic	

Condenser Fan Data (if applicable)

Quantity	
Manufacturer	
Full-Load hp	
Frame Size	
Enclosure Type	
Full-Load rpm	
Efficiency	
Volts	
Phase	
Full-Load Amps	
Service Factor	
Water-Side Free Cooling	yes / no

Condenser Pump Control

Pumps Cycled with Respective Chillers	yes / no
Pumps Operate Independent of Chiller Load	yes / no
Pumps Headered/Chiller Headered	yes / no
Automatic Isolation Valves	yes / no

Provide unusual piping arrangement description/schematic.

Evaporator Pump Control

Pumps Cycled with Respective Chillers	yes / no
Pumps Operate Independent of Chiller Load	yes / no
Pumps Headered/Chillers Headered	yes / no
Automatic Isolation Valves	yes / no

Provide unusual piping arrangement description/schematic.

Controls Notes

Time-of-Day Scheduling ______________________________________

____

Chiller Sequencing ______________________________________

____

2.2 COOLING TOWERS AND FLUID COOLERS

PCBEA Sample Forms

General Information

Site Name			
Tower ID/Tag			
Location/Service			
Manufacturer			
Model Number		Serial Number	
Quantity of Cells		Year Built	

Heat Rejection Type

Cooling Tower:	
Induced Draft	
Forced Draft	
Natural Draft	
Wetted Media	
Fluid Cooler:	
Air Cooled	
Evaporatively Cooled	

Piping Arrangement

Basins Headered	
Piped to Individual Chillers	
Valves on Supply Headers	

Fluid Cooler Pump Data	Design	Measured
Quantity		
Fluid Type		
gpm (each pump)		
Manufacturer		
Full-Load hp		
ODP/TEFC		
Full-Load rpm		
Pump Efficiency		
Motor Efficiency		
Full-Load Amps		
Volts/Phase		
Pump Motor hp		
Power Factor	n/a	

Fan Data

Fan Control	VFD / Single-Speed / Two-Speed
Fan Drive	Belt / Direct / Gear
Fan Type	Centrifugal / Propeller / Other: ______________

Fan Data	Design	Measured
Quantity		
cfm (each fan)		
Full-Load hp		
ODP/TEFC		
Full-Load rpm		
Fan Efficiency		
Motor Efficiency		
Full-Load Amps		
Volts/Phase		
Power Demand (kW)		
Power Factor	n/a	

Condenser Water	Design	Measured
gpm		
EWT/LWT (°F)		
Ambient DB/WB (°F)		

Basin Heater	Design	Measured
Power Demand (kW)		
Full-Load Amps		
Volts/Phase		
Power Factor	n/a	

Controls

Time-of-Day Scheduling ____

____

Tower/Cooler Staging ____

____

Condenser Water Flow Rate ____

____

Fill Condition ____

____

2.3 PUMP AND PIPING SYSTEMS

PCBEA Sample Forms

General Information

Site Name	
Pump ID/Tag	
Location/Service	
Year Built	

Pump Data

Manufacturer	
Model Number	
Serial Number	
Pump Type	
Impeller Size	
Pump Efficiency	

Pump Motor Data

Full-Load hp	
Frame Size	
Enclosure Type	
Full-Load rpm	
Synch. rpm	
Motor Efficiency	
Volts	
Phase	
Full-Load Amps	
Connection Type	

Design/Operating Conditions

	Design	Measured
Total Head (ft. w.g.)		
gpm		
Pump Motor rpm		
Volts		
Amps		
Power Demand (kW)		
Power Factor	n/a	

Pump Control

		Notes
Constant Volume		
Variable Volume		
VFD		
Multi-Speed		
Other		
Connected to DDC?	yes / no	

Pumping Configuration

Primary-Only Variable Flow	
Primary-Only Constant Flow	
Primary/Secondary Variable	
Primary/Secondary/Tertiary	
Other ________________	

Miscellaneous Accessories

Suction Diffusers	
Triple-Duty Valves	
Throttling Valves	
Circuit Setters	
Other ____________	

Controls Notes

Time-of-Day Scheduling

Pump Staging

Balancing Valve Type/Position

2.4 AIR-HANDLING SYSTEM EQUIPMENT

PCBEA Sample Forms

General Information

Site Name			
Air Handler ID/Tag			
Location/Service			
Manufacturer/Model #			
Year Built		Serial Number	

Supply Fan Data

Manufacturer	
Model Number	
Serial Number	
Fan Type/Size	
Fan Efficiency	

Supply Fan Motor Data

Full-Load hp	
Frame Size	
Enclosure Type	
Full-Load rpm	
Synch. rpm	
Motor Efficiency	
Volts	
Phase	
Full-Load Amps	
Motor in Supply Airstream?	yes / no
Drive Type	

Supply Fan Design/Operating Conditions

	Design	Measured
Total Static Pressure		
Occupied cfm		
Unoccupied cfm		
Fan rpm		
Volts		
Occupied Amps		
Unoccupied Amps	n/a	
Occupied kW		
Unoccupied kW	n/a	
Power Factor	n/a	

Filter Data	Pre	Final
Manufacturer		
Model Number		
Type		
Efficiency		
Initial/Final Pressure Drop		

Fan Configuration

Blow-Thru	
Draw-Thru	

Return Air Path

Ducted	
Plenum	

Humidifier

Type	
Capacity (lb/h)	

Cooling Coil	NONE	
Chilled Water (2-Way/3-Way Valve)		
Sensible/Latent Capacity		
Design cfm/gpm		
Design EDB/EWB		
Design LDB/LWB		
Design EWT/LWT		
Measured EWT/LWT		
Air Pressure Drop		

Pre-Heat Coil	NONE	
Medium Type		
Capacity (MBtuh)		
Design cfm/gpm		
Design EDB/LDB		
Design EWT/LWT		
Measured EWT/LWT		

Heating Coil	NONE	
Medium Type		
Capacity (MBtuh)		
Design cfm/gpm		
Design EDB/LDB		
Design EWT/LWT		
Measured EWT/LWT		

2.5 AIR-HANDLING SYSTEM CONTROLS

PCBEA Sample Forms

General Information

Site Name	
Air Handler ID/Tag	
Location/Service	
Manufacturer/Model #	
Year Built	
Serial Number	

Supply Fan Control

Constant Air Volume	
VFD	
Discharge Damper	
Multi-Speed	
DDC	
Electronic Controls	
Pneumatic Controls	
Room Thermostat	

Outdoor Air (OA) Control

Fixed OA cfm	
Economizer?	yes / no
Minimum OA %	
Economizer Control	
Econo Upper Limit	
Econo Lower Limit	
Econo Control Point	

Humidifier Control

Humidity Control?	yes / no
Humidity Sensor Location	
RH Setpoint, Max	
RH Setpoint, Min	

Temperature Control Pump hp

Occupied Room Temperature Setpoint	
Unoccupied Room Temperature Setpoint	
Heating Supply Air Temperature Setpoint	
Cooling Supply Air Temperature Setpoint	
Supply Air Temperature Setpoint Reset	yes / no

Operating Schedule

Occupied Hours	
Unoccupied Hours	
Morning Warm-Up Hours	

Supply Fan Static Pressure Control

Constant Static Pressure Setpoint	
Static Pressure Reset	

Ductwork Control Accessories

Turning Vanes?	
Single/Dual Duct	
Lining?	
Other ________________	

Louver/Damper Control

Location(s)	
% Free Area	
Pressure Drop	
Control Method	

Controls Notes

2.10 PACKAGED UNITS: DX, HEAT PUMPS

PCBEA Sample Forms

General Information

Site Name					
Unit ID/Tag					
Location/Service					
Manufacturer					
Model Number					
Serial Number		Year Built		No. of Zones Served by Unit	

Unit Type

Rooftop Unit	
Split System	
Window A/C	
VAV	
CAV	
Other _________	

Condenser Type

Air-Cooled	
Water-Cooled	
Evap-Cooled	
GSHP	
Loop type________	
WSHP (type______)	
iso valves?	Y / N
Other________	

Preheat/Heating Coil

Water (HWS/HWR) (°F)	
Steam (psi)	
Electric (kW)	
Heating Capacity (MBH)	
Fuel Type______________	

Compressor Data

Quantity	
RLA	
Volts	
Phase	
Type	

Condenser Fan Data

Quantity	
Full-Load Amps	
Volts	
Phase	
Nominal hp	

Supply Fan Motor Data

Quantity	
Full-Load hp	
Frame Size	
Enclosure Type	
Operating rpm	
Motor Efficiency	
ECM ?	Y / N
Volts	
Phase	
Full-Load Amps	
Power Demand (kW)	
Drive Type	

Supply Fan Data

Manufacturer	
Model Number	
Serial Number	
Fan Type/Size	
Fan Efficiency	

Cooling Type

DX: Refrigerant Type	
Number of Circuits	
cfm	
EDB/EWB (°F)	
LDB/LWB (°F)	
Cooling Cap. (MBH)	
EER	
EER incl. Evap. Fan?	Y / N

Humidity Control NONE

Humidifier Type	
Humidity Sensor Location	
RH Setpoint, Max	
RH Setpoint, Min	

Supply Fan Control

VFD	
Discharge Damper	
Multispeed	
DDC	
Pneumatic	
Room Thermostat	

Outdoor Air (OA) Control

Fixed OA cfm	
Economizer?	
Minimum OA %	

Electrical Data

MCA	
MOCP	

Controls Notes

Time-of-Day Scheduling
Unit Staging

Sample Templates

PEA Template

The following instructions and templates are from the spreadsheet "ASHRAE PEA Template.xls." They may be used as a basis for completing a preliminary energy-use analysis (PEA). See Part 1, page 6 for detailed instructions on how to conduct a PEA.

Procedures for Commercial Building Energy Audits, Second Edition

Sample Preliminary Energy-Use Analysis Instructions

Use this template as a starting point for a Preliminary Energy-Use Analysis (PEA).

Link to Sheet	Summary Instructions (See Part 1 of *Procedures for Commercial Building Energy Audits*, Second Edition, for details.)
(1) Gross Floor Area	Determine the building's gross conditioned floor area and record it on the Basic Site Info sheet. Use the standard definition of gross floor area, which is: the sum of the floor areas of all the spaces within the building with no deductions for floor penetrations other than atria. It is measured from the exterior faces of exterior walls or from the centerline of walls separating buildings, but it excludes covered walkways, open roofed-over areas, porches and similar spaces, pipe trenches, exterior terraces or steps, roof overhangs, parking garages, surface parking, and similar features.
(2) Historical Billing	Compile billing information for at least a one-year period, preferably for a two- or three-year period. Summary tables are provided for three years of data for electricity, natural gas, and one other fuel type.
(3) Delivered Energy	Use the Delivered Energy sheet to summarize energy purchases such as propane or fuel oil that are delivered on an irregular basis.
(4) Energy Performance Summary	Complete the Energy Performance Summary to develop the EUI and the ECI for ach fuel or demand type. This template is set up to automatically sum utility consumption and costs for electricity and natural gas from the historical billing data. However, ensure that the totals sum correctly for your site, particularly if you have modified the sheets in any way. Add Delivered Energy totals if applicable.
(5) EUI	Compare the EUI and ECI with those of buildings having similar characteristics. Potential sources include: ENERGY STAR® Target Finder U.S. E.I.A. Commercial Buildings Energy Consumption Survey (CBECS) Energy IQ (Easy interface for CBECS and California specific benchmark data)
(6) EUI	Compare the energy and cost savings for each fuel type if the building were to reach the target EUI (not outlined here). Using these values, determine whether further engineering analysis is recommended.

BASIC SITE INFORMATION

Commercial Building Energy Audit Sample Forms

Project Name

Client Name

Site Address

City/State

Year

Building Type and/or Functions

Gross Area

Lease Type

No. Stories

Key Contacts

Name and Position	Phone/Email

Economic Criteria for Energy Projects

Previous Audit or Engineering Study Availability

Notes

NOTE: Request utility/fuel bills for two- to three-year period.

HISTORICAL UTILITY BILLS

Commercial Building Energy Audit Sample Forms

At least one year of data, or preferably two to three years of data, should be obtained for each meter or energy source in the building. Use multiple sheets as necessary.

ELECTRIC

Utility/Supplier Name	
Account Number	
Meter Number	
Rate Schedule Name/Number	
Notes on Rate	

YEAR:

Month	Bill Date	Days in Period	Ave. Temp. (F)	Actual Demand (kW)	Billed Demand (kW)	Electric Use (kWh)	Demand Cost ($)	Electric Use ($)	Other Fees ($)	Total Bill ($)
January										$0
February										$0
March										$0
April										$0
May										$0
June										$0
July										$0
August										$0
September										$0
October										$0
November										$0
December										$0
					Annual Totals	**0**	**$0**	**$0**	**$0**	**$0**

Peak Demand (kW)	0
Total Annual Cost ($)	$0

HISTORICAL UTILITY BILLS
Commercial Building Energy Audit Sample Forms

At least one year of data, or preferably two to three years of data, should be obtained for each meter or energy source in the building. Use multiple sheets as necessary.

NATURAL GAS

Utility/Supplier Name	
Account Number	
Meter Number	
Rate Schedule Name/Number	
Notes on Rate	

YEAR:

Month	Bill Date	Days in Period	Ave. Temp. (F)	Actual Demand* (______)	Billed Demand* (______)	therms	Demand Cost ($)	Gas Use ($)	Other Fees ($)	Total Bill ($)
January										$0
February										$0
March										$0
April										$0
May										$0
June										$0
July										$0
August										$0
September										$0
October										$0
November										$0
December										$0
					Annual Totals	**0**	**$0**	**$0**	**$0**	**$0**

* Choose appropriate units for gas (typically MMBtu, therms, or MCF).

Peak Demand (______)	0
Total Annual Cost ($)	$0

HISTORICAL UTILITY BILLS

Commercial Building Energy Audit Sample Forms

At least one year of data, or preferably two to three years of data, should be obtained for each meter or energy source in the building. Use multiple sheets as necessary.

OTHER

Utility/Supplier Name	
Account Number	
Meter Number	
Rate Schedule Name/Number	
Notes on Rate	

* Other utilities might be purchased steam, purchased chilled water, LPG, or fuel oil.

YEAR:

Month	Bill Date	Days in Period	Ave Temp (F)	Actual Demand* (______)	Billed Demand* (______)	Use* (______)	Demand Cost ($)	Gas Use ($)	Other Fees ($)	Total Bill ($)
January										$0
February										$0
March										$0
April										$0
May										$0
June										$0
July										$0
August										$0
September										$0
October										$0
November										$0
December										$0
					Annual Totals	**0**	**$0**	**$0**	**$0**	**$0**

* Use units corresponding to those on the bill.

Peak Demand (______)	0
Total Annual Cost ($)	$0

DELIVERED CONSUMPTION DATA

Commercial Building Energy Audit Sample Forms

At least one year of data, or preferably two to three years of data, should be obtained for each delivered service. Get delivered information from the client or the service provider. Print multiple utility sheets as necessary.

Utility Company ____________________

Account Number ____________________

Rate Number ____________________

Energy Type ____________________

Units[1] ____________________

Delivery Date	Delivery Amount	Total Cost[2]
1		
2		
3		
4		
5		
6		
7		
8		
9		
10		
11		
12		
13		
14		
15		
16		
17		
18		
19		
20		
USE OF INVENTORY	C	D
TOTAL CONSUMPTION		

Use of Inventory C

At Date 0 __________ A

365 Days after Date 0 __________ B

Use of Inventory (A-B) 0 C

Value of Inventory Used

Latest Price __________ D

Value (CxD) $0.00

1. gal, lb, etc.

2. Costs should include tax, fees, contract charges, etc.

ENERGY PERFORMANCE SUMMARY

Commercial Building Energy Audit Sample Forms

Energy Type	Total Annual Use	Units	Conversion Multiplier	kBtu	Total Annual Cost ($)
Electricity	0	kWh	3.412142	-	$ -
Natural Gas	0	therms	100	-	$ -
Purchased Steam			0	-	
Purchased Hot Water			0	-	
Purchased Chilled Water			0	-	
Oil #:			0	-	
Propane			0	-	
Coal			0	-	
Thermal—On-Site Generated			0	-	
Other			0	-	
Electricity—On-Site Generated			0	-	
Thermal or Electricity—Exported			0	-	
			0	-	
Total				**-**	**$ -**

Gross Conditioned Area	**0**
EUI (kBtu/ft^2)	**-**
Target Finder Score*	
CBECS EUI (for comparable , kBtu/ft^2)	
ECI ($/ft^2)	**0**

Notes:

* Additional data may be required for your building type.

Sample End-Use Breakdown

The following instructions and templates are from the spreadsheet "ASHRAE Sample End-Use Breakdown.xls." They may be used as a starting point for the end-use cost allocation methods described in Part 2, page 31.

Procedures for Commercial Building Energy Audits, Second Edition
Sample End-Use Breakdown Instructions

Use this template as a starting point for an end-use breakdown (end-use disaggregation).

(1) Estimate the energy consumption of end-use systems in the buildings and enter these estimates into the appropriate column in the table. Spaces for inputs are marked in blue text.

For a Level 1 analysis, these estimates may be based on expected equipment densities or estimates based on similar buildings.

For a Level 2 or Level 3 analysis, the end-use estimates for major end uses may be based upon one of the following:
(listed from most accurate to least accurate)

a) Detailed measurements of end uses, measured over time
b) Spot measurements of end uses, with estimated hours of operation with analysis to support annualization of the measured data
c) Detailed calculations or hourly simulations of equipment performance based on manufacturers' specifications or design drawings
d) Assumptions of typical end-use densities

(2) Input the historical energy consumption for your building.
Replace the inputs in blue text with data for your billing, including the end-use estimates, the building gross floor area (in units of your choosing), and the historical billing data totals for comparison. These inputs should be consistent with the historical energy use used as part of the Preliminary Energy-Use Analysis Template.

(3) Replace the sheet titles with your own. Note project name, date, etc.

(4) Use the space below the table to document your assumptions or to note conclusions that can be derived from the data.

(5) **You may need to add additional end uses for your building, if they are significant.**
For example, if the template were used for a data center, you would want to include a separate line item for input power supplied to uninterruptible power supplies (UPSs). It is the auditor's responsibility to chose end-use categories that are meaningful for the building. These are listed for guidance as typical end-use categories.

(6) Delete rows in the end-use table that are not necessary for your building. This will eliminate blank labels on your chart.

The remaining cells will be calculated as a result. A sample sheet is included for your reference, or you may begin from the blank copy.

Procedures for Commercial Building Energy Audits, Second Edition

Sample End-Use Breakdown

		Conversion Factor to kBtu
Input Unit 1	kWh	3.412142
Input Unit 2	therms	100
Input Unit 3	gallons (propane)	91.33
Combined Output Units	kBtu	1
Building Gross Floor Area	99,999	
Floor Area Units	ft^2	

	Input Energy Units			Combined Energy Use	
End Use	kWh	therms	gallons (propane)	kBtu	%
Air Compressors	25,000	-		85,304	1%
Cooking	36,000	-	9,800	1,017,870	6%
Cooling	445,996	-		1,521,800	10%
Heating	699,993	20,640		4,452,455	28%
Lighting (Exterior)	68,455	-		233,578	1%
Lighting (Interior)	371,996	-		1,269,304	8%
Miscellaneous	-	-	5,600	511,448	3%
Office Equipment	350,856	-		1,197,170	8%
Other Plug Loads	305,997	-		1,044,105	7%
Process	-	27,620		2,761,972	18%
Pumps	56,525	-		192,871	1%
Refrigeration	38,500	-		131,367	1%
Ventilation	146,999	-		501,580	3%
Water Heating	22,000	6,970		772,059	5%
Total Estimated	**2,568,316**	**55,229**	**15,400**	**15,692,885**	**100%**
Historical Billing	**2,575,020**	**56,800**	**15,500**	**14,466,334**	
Percent of Actual	**99.7%**	**97.2%**	**99.4%**	**108.5%**	
Total per ft^2	**25.7**	**0.6**	**0.2**	**156.9**	

Combined Fuel End-Use Breakdown

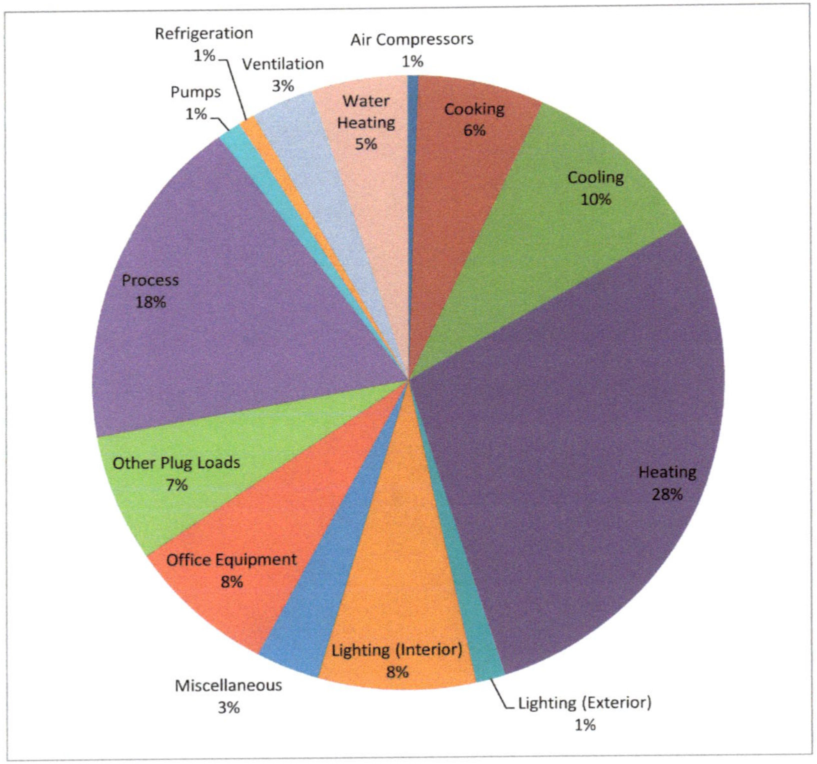

Assumptions / Notes / Conclusions

***Procedures for Commercial Building Energy Audits*, Second Edition**

Sample End-Use Breakdown

End-Use Breakdown Input by kWh

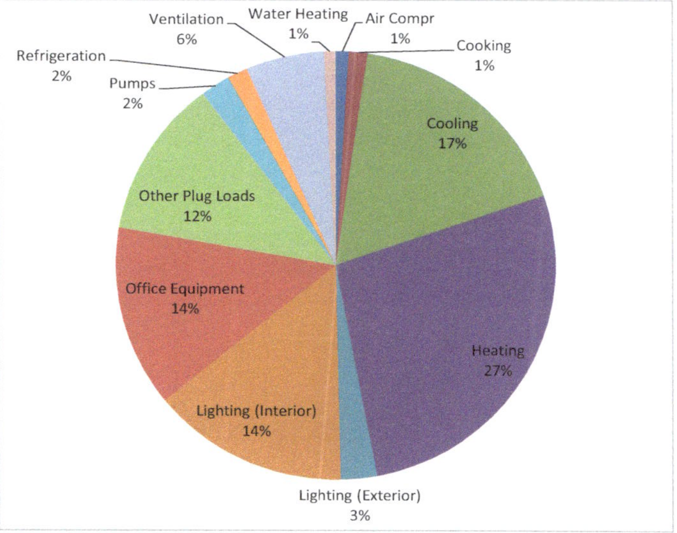

End-Use Breakdown Input by therms

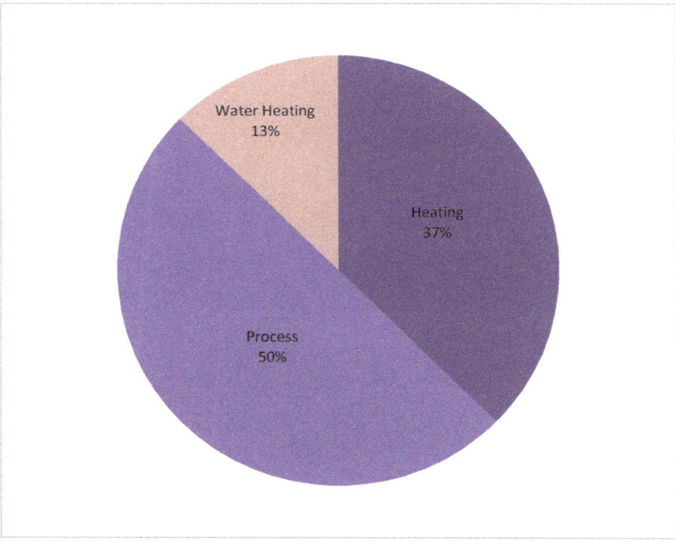

End-Use Breakdown Input by gallons (propane)

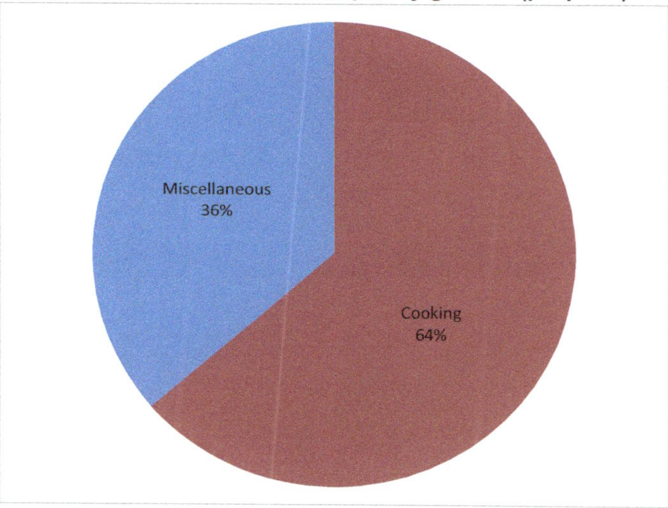

Procedures for Commercial Building Energy Audits, Second Edition

Sample Peak Demand (kW) Breakdown

Peak Demand Units	kW
Building Gross Floor Area	99,999
Floor Area Units	ft^2

End Use	kW	%
Air Compressors	15.5	3%
Cooking	6.5	1%
Cooling	225.0	40%
Heating	-	0%
Lighting (Exterior)	-	0%
Lighting (Interior)	105.0	19%
Miscellaneous	-	0%
Office Equipment	26.0	5%
Other Plug Loads	3.5	1%
Process	1.5	0%
Pumps	35.3	6%
Refrigeration	16.9	3%
Ventilation	123.8	22%
Water Heating	-	0%
Total Estimated	**559.0**	**100%**
Historical Billing	565.3	
Percent of Actual	**98.9%**	
Total per 1000 x ft^2	**5.6**	

Peak Demand (kW) Breakdown

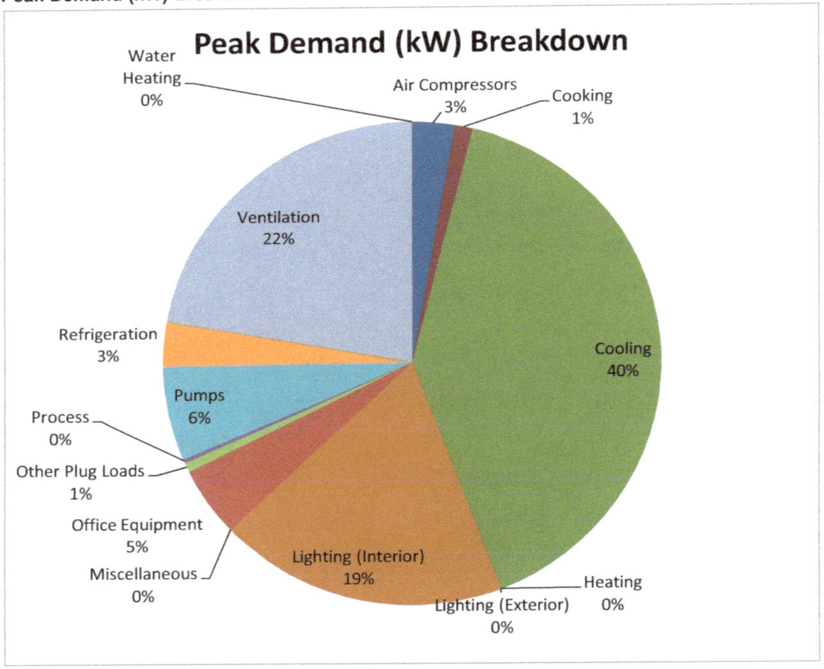

Assumptions / Notes / Conclusions

EEM Summary Table

The following instructions and templates are from the spreadsheet "ASHRAE EEM Summary Table.xls."

Procedures for Commercial Building Energy Audits, Second Edition

Sample EEM Summary Table Instructions

Use this template to report energy efficiency measure (EEM) costs and savings.

(1) Enter the following inputs for each measure (on the spreadsheet, inputs display in blue; all output cells are in black). For each measure, enter the following:

- Measure Description
- Electricity Savings (kWh)
- Gas/Fuel Savings (therms)
- Total Cost Savings
- Measure Cost
- Potential Utility Incentive
- Measure Life

(2) Update the assumption for the real discount rate as appropriate for the customer. A default value of 4% is provided.

(3) Replace the sheet titles with your own. Note project name, date, etc.

(4) Use the space below the table to document your assumptions

(5) You may want to add columns to show more detail such as cost savings associated with demand, consumption, and fuel savings separately.

The remaining cells will be calculated as a result. An example sheet is included for your reference.

All life-cycle cost calculations are consistent with formulas within *Procedures for Commercial Building Energy Audits*, Second Edition.

***Procedures for Commercial Building Energy Audits*, Second Edition**

Sample EEM Summary Table

Measure Number	Measure Description	Annual Energy and Cost Savings				Payback with Incentive						
		Peak Demand Savings (kW)	Electricity Savings (kWh)	Gas/Fuel Savings (therms)	Total Cost Savings	Measure Cost	Potential Utility Incentive	Measure Life (years)	Net Measure Cost	IRR (over Life of Measure)	NPV*	Simple Payback (yr)
EEM-1									$ -	N/A	$ -	N/A
EEM-2									$ -	N/A	$ -	N/A
EEM-3									$ -	N/A	$ -	N/A
EEM-4									$ -	N/A	$ -	N/A
EEM-5									$ -	N/A	$ -	N/A
EEM-6									$ -	N/A	$ -	N/A
EEM-7									$ -	N/A	$ -	N/A
EEM-8									$ -	N/A	$ -	N/A
EEM-9									$ -	N/A	$ -	N/A
EEM-10									$ -	N/A	$ -	N/A
EEM-11									$ -	N/A	$ -	N/A
EEM-12									$ -	N/A	$ -	N/A
TOTALS (Recommended Measures)		-	-	-	$ -	$ -	$ -		$ -	0%	$ -	-

* NPV assumes a discount rate of 4%

(1) The net present value (NPV) is the value (the gain minus the cost) of an investment in today's dollars over some specified time period. If the investment has a positive NPV, it is generally considered to be beneficial.

(2) The internal rate of return (IRR) is the annual yield from a project, usually expressed as a percentage of the total amount invested; the compound rate of interest which, when used to discount cash flows, will result in zero net savings. If the IRR is greater than the investor's stated discount rate, the measure is considered beneficial.

Procedures for Commercial Building Energy Audits, Second Edition

Sample EEM Summary Table

Measure Number	Measure Description	Annual Energy and Cost Savings				Payback with Incentive						
		Peak Demand Savings (kW)	Electricity Savings (kWh)	Gas/Fuel Savings (therms)	Total Cost Savings	Measure Cost	Potential Utility Incentive	Measure Life (years)	Net Measure Cost	IRR (over Life of Measure)	NPV*	Simple Payback (yr)
EEM-1	Replace Incandescent Lamps with CFLs	7.6	15,245	-	$ 1,906	$ 1,875	$ 545	3	$ 1,330	132%	$ 3,958	0.7
EEM-2	Reduce Pressure Setting on Pneumatic Compressor	-	2,312	-	$ 206	$ -	$ -	3	$ -	N/A	$ 571	-
EEM-3	Install VFD on Tenant Condenser Loop Pump to Reduce Flow	19.0	163,872	-	$ 25,188	$ 17,386	$ 13,110	10	$ 4,276	589%	$ 200,021	0.2
EEM-4	Install VFD and Implement Demand-Controlled Ventilation for AHU-1	-	12,448	423	$ 2,290	$ 11,136	$ 1,418	10	$ 9,718	20%	$ 8,855	4.2
EEM-5	Install CO Sensors and VFD to Control Garage Exhaust Fan	18.8	48,948	-	$ 8,811	$ 25,616	$ 3,916	10	$ 21,700	39%	$ 49,762	2.5
EEM-6	Add Hot-Water Resets Control and Install Condensing Boiler	-	44,838	17,203	$ 25,274	$ 102,511	$ 20,790	20	$ 81,721	31%	$ 261,758	3.2
EEM-7	Replace Garage HPS Fixtures with LED Fixtures with Integrated Motion Sensor	6.8	29,854	-	$ 4,114	$ 29,598	$ 2,174	15	$ 27,423	12%	$ 18,316	6.7
EEM-8	Repair Economizers on All Air Handlers	5.0	22,342	-	$ 2,904	$ 12,864	$ 2,511	5	$ 10,354	12%	$ 2,577	3.6
EEM-9									$ -	N/A	$ -	N/A
EEM-10									$ -	N/A	$ -	N/A
EEM-11									$ -	N/A	$ -	N/A
EEM-12									$ -	N/A	$ -	N/A
TOTALS (Recommended Measures)		**57.3**	**339,860**	**17,625**	**$ 70,692**	**$ 200,985**	**$ 44,464**		**$ 156,521**	**44%**	**$ 545,819**	**2.2**

* NPV assumes a discount rate of 4%

(1) The net present value (NPV) is the value (the gain minus the cost) of an investment in today's dollars over some specified time period. If the investment has a positive NPV, it is generally considered to be beneficial.

(2) The internal rate of return (IRR) is the annual yield from a project, usually expressed as a percentage of the total amount invested; the compound rate of interest which, when used to discount cash flows, will result in zero net savings. If the IRR is greater than the investor's stated discount rate, the measure is considered beneficial.

Hourly Simulation Checklist

The following checklist is from the spreadsheet "ASHRAE Hourly Simulation Checklist.xls." Use this brief checklist as a guide to ensure that your model is reasonable and accurate. Note that completing every item on this list will not *guarantee* that your model is accurate and represents your building. Use engineering judgment to investigate any other items that seem aberrant, and pay special attention to unusual systems and components in your building.

Procedures for Commercial Building Energy Audits, Second Edition

Hourly Simulation Checklist

Use this checklist as a guide to ensure that your model is reasonable and accurate. Note that completing every item on this list will not *guarantee* that your model is accurate and represents your building. Use engineering judgment to investigate any items that seem aberrant and pay special attention to unusual systems and components in your building.

Input Verification

Project

- ☐ Verify correct weather file/zone
- ☐ Specify correct site altitude
- ☐ Specify ground reflectance (account for major factors such as nearby lake or mountain ridge)
- ☐ Specify correct utility rate schedules

Envelope

- ☐ Specify walls, windows, floors, and underground floor constructions per drawings
- ☐ Verify constructions per drawings (especially if "wizard" inputs are used)
- ☐ Specify window and frame properties (and check correct treatment of outdoor air coefficient)
- ☐ Verify slab heat transfer coefficients as appropriate
- ☐ Verify building floor area (omit plenums)
- ☐ Include skylights if needed—make sure they are not obscured by roof constructions
- ☐ Include window overhangs where needed
- ☐ Include exterior shading from other parts of building
- ☐ Include exterior shading from other buildings

Spaces

- ☐ Verify equipment internal load density (W/ft^2, W/m^2)
- ☐ Verify equipment schedules
- ☐ Verify lighting loads density (W/ft^2, W/m^2)
- ☐ Verify lighting schedules
- ☐ Verify occupancy loads (number of people or area per person)
- ☐ Verify occupancy schedules
- ☐ Verify infiltration loads (air changes per hour or flow per unit area)
- ☐ Verify infiltration schedules
- ☐ Add daylighting controls as appropriate

Zones

- ☐ Verify zones are consistent with building construction (conditioned, unconditioned, plenums, etc.)
- ☐ Verify zone thermostat setpoints and schedules
- ☐ Set outdoor airflow/person in zones

Air-Side Systems

- ☐ Assign zones to correct systems and specify control zone as appropriate
- ☐ Verify equipment operating schedules
- ☐ Size systems
- ☐ Size heating coil capacities (auto-sizing is discouraged)
- ☐ Size cooling coil capacities (auto-sizing is discouraged)
- ☐ Add exhaust fans and controls as needed
- ☐ Ensure fan energy and compressor power are separate (packaged units)
- ☐ Verify supply air setpoints and resets

Plants

- ☐ Specify configuration of hot-water, condenser-water, and chilled-water loops (You may need separate loops for primary and secondary loops)
- ☐ Verify equipment operating schedules, capacities, and part-load efficiencies
    - ☐ Boiler(s) (if multiple, ensure correct staging)
    - ☐ Chiller(s) (if multiple, ensure correct staging)
    - ☐ Pump(s) (if multiple, ensure correct staging)
- ☐ Verify equipment control sequences
- ☐ Check pump definitions use (1) kW input or (2) flow rate, pressure drop, and efficiency (not both)

Procedures for Commercial Building Energy Audits, Second Edition

Hourly Simulation Checklist

Output Verification

Review building performance summary reports

- ☐ Check overall EUI (compare with known benchmarks)
- ☐ Check consistency with historical utility billing data (all end uses)
- ☐ Check end-use splits for reasonableness (compare with known benchmarks)
- ☐ Check end-use splits for reasonableness (compare with known benchmarks or sub-metered data)
    - ☐ Equipment/plug loads
    - ☐ Lighting
    - ☐ Final occupant density
- ☐ Check final equipment design/sizing
    - ☐ Cooling capacity
    - ☐ Heating capacity
    - ☐ Airflow
    - ☐ Loop sizing, flow rates
    - ☐ Ventilation design flow rate
- ☐ Check final system summary performance
    - ☐ Heating average system efficiency
    - ☐ Cooling average system efficiency

Review detailed hourly reports and check loads are met (zones and plant)

- ☐ Check loads are met (zones and plant)
- ☐ Verify hourly equipment schedules

Additional Resources

Other forms, templates, checklists, and lists are available for download from www.ashrae.org/PCBEA, including the following:

- A pdf of an energy auditor checklist redistributed with permission from Washington State University
- A pdf of a list of EEMs to consider when performing an energy audit. The list is organized by end use and building component and includes some water efficiency measures. The list was developed by ASHRAE Standard Project Committee 100 in 2011 as part of the committee's rewrite of ANSI/ASHRAE Standard 100-2006.

Unit Conversions

The following tables provide information on converting Inch-Pound (I-P) units to Systéme International (SI) units. This information is also available at www.ashrae.org/PCBEA in the Formulas and Unit Conversions spreadsheet.

Energy Units—kBtu Conversions

Multiply the baseline quantity by the conversion multiplier to obtain a result in kBtu.

Fuel	Measured Units	Conversion Multiplier	Source
Any	MMBtu	1000	
	therms	100	
	dekatherms	1000	
Electricity	kWh	3.412142	
	MWh	3412.142	
Natural Gas	ft^3	1.027	
	CCF (100 x ft^3)	102.7	2
	MCF (1000 x ft^3)	1027	2
	therm	100	2
	m^3	36.4	
	MJ	0.9478171	
	GJ	947.8171	
Purchased Steam	1000 Btu	1	
	1000 lb (approx)	970	
	lb (steam, 15 psig, evap)	0.945684	3
	lb (steam, 50 psig, evap)	0.912061	3
	therm	100	
Purchased Hot Water	1000 Btu	1	
Purchased Chilled Water	1000 Btu	1	
	ton-hour	12	
Oil #1	U.S. gallon	137.4	1
Oil #2	U.S. gallon	139.6	1
	Imperial gallon	167	
	litre	36.7	

Energy Units—kBtu Conversions *(continued)*

Multiply the baseline quantity by the conversion multiplier to obtain a result in kBtu.

Fuel	Measured Units	Conversion Multiplier	Source
Oil #3	U.S. gallon	141.8	1
Oil #4	U.S. gallon	145.1	1
Oil #5	U.S. gallon	148.8	1
Oil #6	U.S. gallon	152.4	1
	Imperial gallon	185	
	litre	40.7	
Diesel	U.S. gallon	139	5
Gasoline	U.S. gallon	124	6
Propane	U.S. gallon	91.33	2
	cubic feet	2.55	3
	Imperial gallon	110	
	litre	24.2	
Anthracite Coal	ton	25,400	
	short ton (coal)	19,953	4

Sources:
1 ASHRAE (2009), Table 6, page 28.7.
2 ASHRAE (2009), page 28.5.
3 ASHRAE (2009), Refrigerant-718 (steam) table, page 30.37.
(Note this is heat of vaporization only.)
4 ASTM (2005).
5 ASTM (2011a).
6 ASTM (2011b).

Formulas and Properties

Formulas		Notes
Sensible Air Conditioning	Btu/h = 1.08 * ΔT * cfm	
Latent Air Conditioning	Btu/h = 4.5 * cfm * ΔH	
Water Heating/Cooling	Btu/h = 500 * gpm * ΔT	
	Btu/h = gpm * ΔT * 8.34 Btu/gal-ΔT * 60 min/h * 1 kW/1000	Derivation
	tons = gpm * ΔT/24	
Water hp	= ft head * gpm/3960	
Pump Motor hp	= water hp/(pump eff * motor eff)	
Pump Hydraulic hp	= gpm * (ft H_2O)/3960	
Fan Motor hp	= cfm * SP/(6354 * Motor Eff * Fan Eff)	6354 cfm * in. H_2O/hp
kW	= 0.746 * hp * (Load Factor)/(Motor Eff)	
EER	= (Btu/h of Cooling Delivered)/(W Input)	
	= (kBtu of Cooling Delivered)/(kWh Input)	
HSPF	= (kBtu of Heating Delivered)/(kWh Input)	Seasonal Average
COP	= EER/3.412	
	= 3.516/(kW/ton)	
	= 1/EIR	
kW/ton	= 12/EER	
EIR	= 1/COP	
	= (kW/ton)/3.516	
Other Misc Unit Equivalents		
1 hp	0.7457 kW	
1 ton	12,000 Btu/h	
1 lb H_2O	7000 gr	
1 psi	2.307 ft H_2O	
1 U.S. gallon	0.1337 ft^3	
1 Imperial gallon	0.1605 ft^3	
1 ft^3 H_2O	62.4 lbm H_2O	
Material	**Properties**	
Steam	Enthalpy of Vaporization	970 Btu/lb
Water	Density	8.34 lbm/gallon
	Specific Heat	1 Btu/(lbm·°F)
Propane	Heat of Combustion	2500 Btu/ft^3
	Heat of Combustion	21,560 Btu/lb
Natural Gas	Heat of Combustion	1000 Btu/ft^3
Wood	Heat of Combustion	8000 Btu/lb
	Density	45 lb/ft^3

Abbreviations and Acronyms

AEE	Association of Energy Engineers
AIRR	adjusted internal rate of return
ASHRAE	American Society of Heating, Refrigerating and Air-Conditioning Engineers
BAS	building automation system
BEAP	Building Energy Assessment Professional certification from ASHRAE
BLCC	Building Life Cycle Cost
BMS	building management system
Btu	British thermal unit
CBECS	Commercial Buildings Energy Consumption Survey
CMVP	Certified Measurement and Verification Professional certification from AEE and Efficiency Valuation Organization
COP	coefficient of performance
Cx	commissioning
$d_{nominal}$	nominal discount rate
d_{real}	real discount rate
DDC	direct digital control
DHW	domestic hot water
DOE	U.S. Department of Energy
ECI	Energy Cost Index
ECM	energy conservation measure
EEM	energy efficiency measure
EERE	Office of Energy Efficiency and Renewable Energy
EIA	Energy Information Administration
EMS	energy management system
EPA	U.S. Environmental Protection Agency
ESCO	energy service company
EUI	Energy Utilization Index
ft	foot
HBDP	High-Performance Building Design Professional certification from ASHRAE
HVAC	heating, ventilating, and air-conditioning
i	inflation rate
IAQ	indoor air quality
IES	Illuminating Engineering Society of North America

IEQ	indoor environmental quality
IRR	internal rate of return
IRS	Internal Revenue Service
KVAR	kilovolt-ampere reactive
kW	kilowatt
kWh	kilowatt-hour
LCC	life-cycle cost
LCCA	life-cycle cost analysis
m	meter
M&E	mechanical and electrical
M&V	measurement and verification
MJ	megajoule
NIST	National Institute of Standards and Technology
NOAA	National Oceanic and Atmospheric Administration
NPV	net present value
O&M	operation and maintenance
OSHA	Occupational Safety and Health Administration
PEA	Preliminary Energy-Use Analysis
PPE	personal protection equipment
QC	quality control
RMI	Rocky Mountain Institute
SIR	savings to investment ratio
SPP	simple payback period
SROI	simple return on investment
UPS	uninterruptible power supply
VAV	variable air volume
VFD	variable-frequency drive

References

Abushakra, B., A. Sreshthaputra, J.S. Haberl, and D.E. Claridge. 2001. Compilation of diversity factors and schedules for energy and cooling load calculations. ASHRAE Research Project 1093 Final Report. Peachtree Corners, GA: ASHRAE.

ASHRAE. n.d. Measuring, Modeling, Analysis, and Reporting Protocols for Short-Term M&V of Whole Building Energy. ASHRAE Research Project 1404. Peachtree Corners, GA: ASHRAE.

ASHRAE. 2001. ANSI/ASHRAE/IES Standard 90.1-2001, *Energy Standard for Buildings Except Low-Rise Residential Buildings*. Peachtree Corners, GA: ASHRAE.

ASHRAE. 2002. ASHRAE Guideline 14, *Measurement of Energy and Demand Savings*. Peachtree Corners, GA: ASHRAE.

ASHRAE. 2007a. ANSI/ASHRAE Standard 105, *Standard Methods of Measuring, Expressing, and Comparing Building Energy Performance*. Peachtree Corners, GA: ASHRAE.

ASHRAE. 2007b. *ASHRAE Handbook—HVAC Applications*. Peachtree Corners, GA: ASHRAE.

ASHRAE. 2009. *ASHRAE Handbook—Fundamentals*. Peachtree Corners, GA: ASHRAE.

ASHRAE. 2010a. *Performance Measurement Protocols for Commercial Buildings*. Peachtree Corners, GA: ASHRAE.

ASHRAE. 2010b. ANSI/ASHRAE/IES Standard 90.1-2010, *Energy Standard for Buildings Except Low-Rise Residential Buildings*. Peachtree Corners, GA: ASHRAE.

ASTM. 2005. ASTM Standard D388, *Standard Classification of Coals by Rank*. West Conshohocken, PA: ASTM International.

ASTM. 2011a. ASTM Standard D975, *Standard Specification for Diesel Fuel Oils*. West Conshohocken, PA: ASTM International.

ASTM. 2011b. ASTM Standard D4814, *Standard Specification for Automotive Spark-Ignition Engine Fuel*. West Conshohocken, PA: ASTM International.

CCC. 2008. Energy Charting and Metrics (ECAM) Tool. www.cacx.org/resources/rcxtools/spreadsheet_tools.html. CA: California Commissioning Collaborative.

CCC. 2010. Building Optimization Analysis (BOA) Tool. www.cacx.org/resources/rcxtools/. CA: California Commissioning Collaborative.

CEC. 2008. Title 24, Part 6, *California Code of Regulations*. Sacramento, CA: California Energy Commission.

EERE. 2002. *International Performance Measurement and Verification Protocol.* Washington, DC: U.S. Department of Energy, Office of Energy Efficiency and Renewable Energy.

EERE. 2011a. EnergyPlus Energy Simulation Software. http://apps1.eere.energy.gov/buildings/energyplus/. Washington, DC: U.S. Department of Energy, Office of Energy Efficiency and Renewable Energy.

EERE. 2011b. AirMaster+. www1.eere.energy.gov/industry/bestpractices/ software_airmaster.html. Washington, DC: U.S. Department of Energy, Office of Energy Efficiency and Renewable Energy.

EERE. 2011c. Building Energy Software Tools Directory. http://apps1.eere.energy.gov/buildings/tools_directory/. Washington, DC: U.S. Department of Energy, Office of Energy Efficiency and Renewable Energy.

EIA. 2003. Commercial Buildings Energy Consumption Survey. www.eia.doe.gov/emeu/cbecs. Washington, DC: U.S. Energy Information Agency.

EIA. 2011. Commercial Buildings Energy Consumption Survey (CBECS), www.eia.doe.gov/emeu/cbecs. Washington, DC: U.S. Department of Energy, U.S. Energy Information Administration.

EPA. 2011a. Portfolio Manager Overview. ENERGY STAR Portfolio Manager. www.energystar.gov/index.cfm?c=evaluate_performance.bus_portfolio manager. Washington, DC: U.S. Environmental Protection Agency.

EPA. 2011b. ENERGY STAR. www.energystar.gov.Washington, DC: U.S. Environmental Protection Agency.

FEMP. 2008. *M&V Guidelines: Measurement and Verification for Federal Energy Projects*, Version 3.0. Washington, DC: U.S. Department of Energy, Office of Energy Efficiency and Renewable Energy, Federal Energy Management Program.

GPO. 2005. *Energy Policy Act of 2005*, Public Law 109-58 (August 8, 2005). Washington, DC: Government Printing Office.

IEA. 2009. *IEA ECBCS Annex 46: Energy & Process Assessment Protocol.* Paris: International Energy Agency, Energy Conservation in Buildings and Community Systems Programme.

JJH. 2009a. DOE-2 Building Energy Use and Cost Analysis Tool. http://doe2.com/DOE2/index.html. Camarillo, CA: James J. Hirsch.

JJH. 2009b. eQUEST—The QUick Energy Simulation Tool. http://doe2.com/equest/. Camarillo, CA: James J. Hirsch.

Kirk, S.J., and A.J. Dell'Isola. 1995. *Life Cycle Costing for Design Professionals*, 2d ed. New York: McGraw-Hill.

Kissock, J.K., J.S. Haberl, and D.E. Claridge. 2002. Development of a toolkit for calculating linear, change-point linear and multiple-linear inverse building energy analysis models. ASHRAE Research Project 1050 Final Report. Peachtree Corners, GA: ASHRAE.

LBNL. 2011. EnergyIQ: Action-Oriented Energy Benchmarking. www.energyiq.lbl.gov. Berkeley, CA: Lawrence Berkeley National Laboratory.

NIST. 2010. Building Life Cycle Cost (BLCC) Program. www1.eere.energy.gov/femp/information/download_blcc.html. Gaithersburg, MD: National Institute of Standards and Technology.

NIST. 1995. NIST Handbook 135, *Life-Cycle Costing Manual for the Federal Energy Management Program*. Gaithersburg, MD: National Institute of Standards and Technology.

Reddy, T.A., and I. Maor. 2006. Procedures for Reconciling Computer-Calculated Results with Measured Energy Data. ASHRAE Research Project 1051 Final Report. Peachtree Corners, GA: ASHRAE.

RSMeans. 2011. RSMeans. http://rsmeans.reedconstructiondata.com. Norwell, MA: Reed Construction Data.

UTOnline.org. 2009. Universal Translator 2, Ver 2.5.115. http://utonline.org/cms/node/54. San Francisco, CA: Pacific Energy Center, UTOnline.org.